全国高职高专工业机器人专业"十三五"规划系列教材
工业机器人应用型人才培养指定用书

工业机器人专业英语

主　编　张明文
副主编　霰学会　宁　金
参　编　顾三鸿　张广才
主　审　高文婷　王　伟

华中科技大学出版社
中国·武汉

内 容 简 介

本书是针对工业机器人应用人才培养的需要而编写的,以培养和提高读者机器人专业英语能力为目标,旨在使读者掌握工业机器人专业英语知识及应用现状。本书由12个主题单元组成,内容涵盖了工业机器人领域的主要技术分支,包括基本知识,不同机器人类型,ABB、KUKA、YASKAWA、FANUC等主流机器人,工业机器人在搬运、焊接、喷涂、装配、打磨等行业的应用,新型机器人,全球机器人发展计划,工业机器人展望等内容。

本书取材新颖、内容丰富,可作为工业机器人等相关专业的专业英语教材,也可供从事机器人相关领域工作的技术人员参考使用。

本书配套有丰富的教学资源,凡使用本书作为教材的教师可咨询相关机器人实训装备,也可通过书末"教学资源获取单"获取相关数字教学资源。咨询邮箱:edubot_zhang@126.com。

图书在版编目(CIP)数据

工业机器人专业英语/张明文主编. —武汉:华中科技大学出版社,2017.9(2022.7重印)
全国高职高专工业机器人专业"十三五"规划系列教材
ISBN 978-7-5680-3262-9

Ⅰ.①工…　Ⅱ.①张…　Ⅲ.①工业机器人-英语-高等职业教育-教材　Ⅳ.①TP242.2

中国版本图书馆 CIP 数据核字(2017)第 187208 号

工业机器人专业英语　　　　　　　　　　　　　　　　　张明文　主编
Gongye Jiqiren Zhuanye Yingyu

策划编辑:万亚军　霰学会
责任编辑:程　青　顾三鸿
封面设计:肖　婧
责任校对:李　琴
责任监印:周治超
出版发行:华中科技大学出版社(中国·武汉)　　电话:(027)81321913
　　　　武汉市东湖新技术开发区华工科技园　　　邮编:430223
录　　排:武汉三月禾文化传播有限公司
印　　刷:武汉开心印印刷有限公司
开　　本:787mm×1092mm　1/16
印　　张:12.5
字　　数:312千字
版　　次:2022年7月第1版第9次印刷
定　　价:39.80元

本书若有印装质量问题,请向出版社营销中心调换
全国免费服务热线:400-6679-118　竭诚为您服务
版权所有　侵权必究

全国高职高专工业机器人专业"十三五"规划系列教材

编审委员会

名誉主任 蔡鹤皋

主　　任 韩杰才　李瑞峰　付宜利

副 主 任 于振中　张明文

委　　员（按姓氏首字母排序）

包春红　陈欢伟　陈健健　陈　霞　董　璐
封佳诚　高春能　高文婷　顾德仁　顾三鸿
韩国震　韩　青　何定阳　赫英强　华成宇
开　伟　李　闻　刘馨芳　卢　昊　宁　金
齐建家　孙锦全　邰文涛　滕　武　王东兴
王璐欢　王　伟　王伟夏　王　艳　吴冠伟
吴战国　夏　秋　霰学会　杨浩成　杨润贤
姚立波　殷召宝　尹　政　喻　杰　张广才
章　平　郑宇琛

序一

现阶段,我国制造业面临资源短缺、劳动力成本上升、人口红利减少等压力,而工业机器人的应用与推广,将极大地提高生产效率和产品质量,降低生产成本和资源消耗,有效提高我国制造业竞争力。我国《机器人产业发展规划(2016—2020年)》强调,机器人是先进制造业的关键支撑装备,也是未来生活方式的重要切入点。广泛采用工业机器人,对促进我国先进制造业的崛起有着十分重要的意义。"机器换人,人用机器"的新型制造方式有效推进了工业升级和转型。

工业机器人作为集众多先进技术于一体的现代制造业装备,自诞生起至今已经取得了长足进步。当前,新科技革命和产业变革正在兴起,全球工业竞争格局面临重塑,世界各国及国际经济组织紧抓历史机遇,纷纷出台相关战略:美国的"再工业化"战略、德国的"工业4.0"计划、欧盟的"2020增长战略",以及我国推出的"中国制造2025"战略。这些战略都以发展先进制造业为重点,并将机器人作为智能制造的核心发展方向。伴随机器人技术的快速发展,工业机器人已成为柔性制造系统(FMS)、工厂自动化(FA)系统、计算机集成制造系统(CIMS)等先进制造系统的关键支撑装备。

随着工业化和信息化的快速推进,我国工业机器人市场进入高速发展时期。国际机器人联合会(IFR)的统计数据显示,截至2016年,中国已成为全球最大的工业机器人市场。未来几年,中国工业机器人市场仍将保持高速的增长态势。然而,现阶段我国机器人技术人才匮乏,与巨大的市场需求严重不协调。《中国制造2025》强调要健全、完善中国制造业人才培养体系,为推动中国从制造业大国向制造业强国转变提供人才保障。从国家战略层面而言,为推进智能制造的产业化发展,工业机器人技术人才的培养刻不容缓。

目前,随着"中国制造2025"战略的全面实施和国家职业教育改革的发展,许多应用型本科院校、职业院校和技工院校纷纷开设工业机器人相关专业。但工业机器人是一门涉及知识面很广的实用型学科,就该学科而言,各院校普遍存在师资力量缺乏、配套教材资源不完善、工业机器人实训装备不系统、技能考核体系不完善等问题,导致无法培养出企业需要的专业机器人技术人才,从而严重制约了我国机器人技术的推广和智能制造业的发展。江苏哈工海渡工业机器人有限公司依托哈尔滨工业大学在机器人方向的研究实力,顺应形势需要,将产、学、研、用相结合,组织企业专家和一线科研人员开展了一系列企业调研,面向企业需求,联合多所高校教师共同编写了"全国高职高专工业机器人专业'十三五'规划系列教材"。

该系列教材具有以下特点:

(1) 循序渐进,系统性强。该系列教材涵盖了工业机器人的入门实用、技术基础、实训指导、工业机器人的编程与高级应用等内容,由浅入深,有助于学生系统地学习工业机器人技术。

（2）配套资源丰富多样。该系列教材配有相应的电子课件、视频等教学资源，并且可提供配套的工业机器人教学装备，构建了立体化的工业机器人教学体系。

（3）通俗易懂，实用性强。该系列教材言简意赅、图文并茂，既可用于应用型本科院校、职业院校和技工院校的工业机器人应用型人才培养，也可供从事工业机器人操作、编程、运行、维护与管理等工作的技术人员参考和学习。

（4）覆盖面广，应用广泛。该系列教材介绍了国内外主流品牌机器人的编程、应用等相关内容，顺应国内机器人产业人才发展需要，符合制造业人才发展规划。

"全国高职高专工业机器人专业'十三五'规划系列教材"结合实际应用，将教、学、用有机结合，有助于读者系统学习工业机器人技术和强化提高实践能力。本系列教材的出版发行，必将提升我国工业机器人相关专业的教学效果，全面促进"中国制造2025"战略下我国工业机器人技术人才的培养和发展，大力推进我国智能制造产业变革。

中国工程院院士 蔡鹤皋

2017年6月于哈尔滨工业大学

序二

自机器人出现至今短短几十年中,机器人技术的发展取得了长足进步,伴随着产业变革的兴起和全球工业竞争格局的全面重塑,机器人产业发展越来越受到世界各国的高度关注,其纷纷将发展机器人产业上升到国家战略层面,提出"以先进制造业为重点战略,以机器人为核心发展方向",并将此作为保持和重获制造业竞争优势的重要手段。

工业机器人作为目前技术发展最成熟且应用最广泛的一类机器人,已广泛应用于汽车及其零部件制造,电子、机械加工,模具生产等行业以实现自动化生产,并参与到了焊接、装配、搬运、打磨、抛光、注塑等生产制造过程之中。工业机器人的应用,既有利于保证产品质量、提高生产效率,又可避免大量工伤事故,有效推动了企业和社会生产力的发展。作为先进制造业的关键支撑装备,工业机器人影响着人类生活和经济发展的方方面面,已成为衡量一个国家科技创新和高端制造业水平的重要标志。

随着工业大国相继提出机器人产业策略,如德国的"工业4.0"、美国的"先进制造伙伴计划"、中国的"'十三五'规划"与"中国制造2025"等国家政策,工业机器人产业迎来了快速发展态势。随着劳动力成本上涨、人口红利逐渐消失,生产方式向柔性、智能、精细化方向转变,中国制造业正处于转型升级的关键时间。在全球新科技革命和产业变革与中国制造业转型升级形成历史性交汇的这一时期,中国成为全球最大的工业机器人市场。大力发展工业机器人产业,对于打造我国制造业新优势、推动工业转型升级、加快制造强国建设、改善人民生活水平具有深远意义。

我国工业机器人产业迎来了爆发性的发展机遇,然而,现阶段我国工业机器人领域人才储备数量严重不足,从工业机器人的基础操作维护人员到高端技术人才普遍存在巨大缺口,企业缺乏受过系统培训、能熟练安全应用工业机器人的专业人才。现代工业是立国的基础,需要有与时俱进的职业教育和人才培养配套资源。"全国高职高专工业机器人专业'十三五'规划系列教材"由江苏哈工海渡工业机器人有限公司联合众多高校和企业共同编写完成。该系列教材依托哈尔滨工业大学的先进机器人研究技术而编写,结合企业实际用人需求,充分贯彻了现代应用型人才培养"淡化理论,技能培养,重在运用"的指导思想。该系列教材涵盖了国际主流品牌和国内主要品牌机器人的入门实用、实训指导、技术基础、高级编程等内容,注重循序渐进与系统学习,并注重强化学生的工业机器人专业技术能力和实践操作能力,既可作为工业机器人技术或机器人工程专业的教材,也可作为机电一体化、自动化专业所开设的工业机器人相关课程的教学用书。

该系列教材"立足工业,面向教育",填补了我国在工业机器人基础应用及高级应用系列教材中的空白,有助于推进我国工业机器人技术人才的培养和发展,助力"中国智造"。

中国科学院院士 韩杰才

2017年6月

前　　言

机器人是先进制造业的重要支撑装备，也是未来智能制造业的关键切入点，工业机器人作为机器人家族中的重要一员，是目前技术最成熟、应用最广泛的一类机器人。工业机器人的研发和产业化应用是衡量一个国家科技创新和高端制造发展水平的重要标志。发达国家已经把发展工业机器人产业作为抢占未来制造业市场、提升竞争力的重要途径，汽车工业、电子电器行业、工程机械等众多行业大量使用工业机器人自动化生产线，在保证产品质量的同时，改善了工作环境，提高了社会生产效率，有力推动了企业和社会生产力发展。

当前，随着我国劳动力成本上涨、人口红利逐渐消失，生产方式向柔性、智能、精细化方向转变，构建新型智能制造体系迫在眉睫，对工业机器人的需求大幅增长。大力发展工业机器人产业，对于打造我国制造业新优势，推动工业转型升级，加快制造强国建设，改善人民生活水平具有深远意义。"中国制造2025"将机器人列入了十大重点发展领域，机器人产业已经上升到了国家战略层面。

在全球范围内的制造产业战略转型期，我国工业机器人产业迎来革命性的发展机遇。然而，现阶段我国工业机器人领域人才供需失衡，缺乏受过系统培训、能熟练安全使用和维护工业机器人的专业人才。针对这一现状，为了更好地推广工业机器人技术的运用，亟须编写一套系统、全面的工业机器人实用教材。

本书是针对工业机器人应用人才培养的需要而编写的，以培养和提高读者机器人专业英语能力为目标，旨在使读者掌握工业机器人专业英语知识及应用现状。本书由12个主题单元组成，内容涵盖了工业机器人领域的主要技术分支，包括基本知识，不同机器人类型，ABB、KUKA、YASKAWA、FANUC等主流机器人，工业机器人在搬运、焊接、喷涂、装配、打磨等行业的应用，新型机器人，全球机器人发展计划，工业机器人展望等内容。

本书取材新颖、内容丰富，可作为工业机器人等相关专业的专业英语教材，也可供从事机器人相关领域工作的技术人员参考使用。

本书由哈工海渡机器人学院的张明文任主编，霰学会和宁金任副主编，参加编写的还有顾三鸿和张广才等，由高文婷和王伟主审。全书由张明文统稿，具体编写分工如下：霰学会编写第1~3章；宁金编写第4~6章；顾三鸿编写第7~9章；张广才编写第10~12章。本书编写过程中，得到了哈工大机器人集团和江苏哈工海渡工业机器人有限公司的有关领导、工程技术人员，以及哈尔滨工业大学相关教师的鼎力支持与帮助，在此表示衷心的感谢！

由于编者水平及时间有限，书中难免有不足之处，敬请读者批评指正。

编　者

2017年6月

目 录

Unit 1　Introduction of Robot …………………………………………………………… (1)
   Part 1　About Robot ……………………………………………………………… (1)
   Part 2　Types and Application of Robot ……………………………………… (4)
   Part 3　About Industrial Robot ………………………………………………… (6)
   Part 4　Component and Applications of Industrial Robot …………………… (9)
Unit 2　Introduction of Industrial Robot …………………………………………… (13)
   Part 1　Main Parts of Industrial Robot ……………………………………… (13)
   Part 2　Basic Terms …………………………………………………………… (15)
   Part 3　Technical Parameters ………………………………………………… (18)
   Part 4　 Kinematics and Dynamics …………………………………………… (22)
Unit 3　Types of Industrial Robots …………………………………………………… (25)
   Part 1　Cartesian Coordinate Robot ………………………………………… (25)
   Part 2　SCARA Robot ………………………………………………………… (27)
   Part 3　Six-axis Articulated Robot …………………………………………… (29)
   Part 4　Palletizing Robot ……………………………………………………… (31)
   Part 5　Delta Robot …………………………………………………………… (33)
Unit 4　ABB Robot ……………………………………………………………………… (35)
   Part 1　ABB and ABB Robot ………………………………………………… (35)
   Part 2　ABB Product Series …………………………………………………… (38)
   Part 3　The Structure of ABB Robot ………………………………………… (41)
   Part 4　Typical Robot—IRB 120 ……………………………………………… (44)
Unit 5　KUKA Robot …………………………………………………………………… (48)
   Part 1　KUKA and KUKA Robot ……………………………………………… (48)
   Part 2　KUKA Robot Product Series ………………………………………… (51)
   Part 3　The Structure of KUKA Robot ……………………………………… (54)
   Part 4　Typical Robot—KR 6 R700 …………………………………………… (57)
Unit 6　YASKAWA Robot ……………………………………………………………… (61)
   Part 1　YASKAWA and YASKAWA Robot …………………………………… (61)
   Part 2　YASKAWA Robot Product Series …………………………………… (63)
   Part 3　The Structure of YASKAWA Robot ………………………………… (66)
   Part 4　Typical Robot—MH12 ………………………………………………… (70)

Unit 7　FANUC Robot ……………………………………………………………… (73)
　　Part 1　FANUC and FANUC Robot …………………………………………… (73)
　　Part 2　FANUC Robot Product Series ………………………………………… (75)
　　Part 3　The Structure of FANUC Robot ……………………………………… (78)
　　Part 4　Typical Robot—LR Mate 200iD/4S …………………………………… (81)

Unit 8　SCARA Robot ……………………………………………………………… (84)
　　Part 1　EPSON SCARA Robot ………………………………………………… (84)
　　Part 2　YAMAHA SCARA Robot ……………………………………………… (87)
　　Part 3　The Structure of SCARA Robot ……………………………………… (90)
　　Part 4　Applications of SCARA Robot ………………………………………… (93)

Unit 9　Industry Application of Robot …………………………………………… (96)
　　Part 1　Painting Robot ………………………………………………………… (96)
　　Part 2　Welding Robot ………………………………………………………… (99)
　　Part 3　Handling Robot ………………………………………………………… (101)
　　Part 4　Assembly Robot ………………………………………………………… (103)
　　Part 5　Polishing Robot ………………………………………………………… (106)

Unit 10　New Types of Robots …………………………………………………… (109)
　　Part 1　YUMI …………………………………………………………………… (109)
　　Part 2　SDA10F ………………………………………………………………… (112)
　　Part 3　Baxter …………………………………………………………………… (114)
　　Part 4　YouBot ………………………………………………………………… (117)
　　Part 5　NAO …………………………………………………………………… (119)

Unit 11　The Outlook for Industrial Robot ……………………………………… (121)
　　Part 1　Current Situation of Industrial Robot ………………………………… (121)
　　Part 2　Development Trend of Robot ………………………………………… (125)
　　Part 3　Social Issues Caused by Application of Robot ……………………… (128)
　　Part 4　Latest Industry Data of Industrial Robot …………………………… (131)

Unit 12　Intelligent Manufacturing and Global Robot Development Program ……… (134)
　　Part 1　Industry 4.0 …………………………………………………………… (134)
　　Part 2　Core Technology of Intelligent Manufacturing ……………………… (137)
　　Part 3　Wisdom Factory ………………………………………………………… (140)

附录1　参考译文 ……………………………………………………………………… (142)
附录2　常用缩略词 …………………………………………………………………… (184)
参考文献 ………………………………………………………………………………… (186)

Unit 1　Introduction of Robot

Robots are used everywhere not only in working places but also at home. These machines can do work that is too dangerous for humans, help around the house, or can be used just for fun.

Part 1　About Robot

What are Robots and Robotics?

The word "robot" comes from the play R. U. R. (Rossum's Universal Robots) written by Karel Čapek in 1920, which means "forced work or labor." The play began in a factory that used robots. These robots were described as efficient but emotionless, incapable of original thinking and indifferent to self-preservation, as shown in Figure 1-1. This was the earliest idea of industrial robots.

Figure 1-1　A Scene from the Play R. U. R.

Today, robot means any man-made machine that can perform work or other actions normally performed by humans, either automatically or by remote control. Robots are machines that can be used to do jobs. Some robots can do work by themselves. Other robots must always be told what to do.

Robotics is a branch of technologies that deals with the study, design and use of robot systems. Robotics is related to the sciences of electronics, engineering, mechanics, and software. These technologies deal with robots that can take the place of humans in dangerous environments or manufacturing processes, or imitate humans in appearance, behaviour, and cognition. Today, many ideas about robots are inspired by nature, contributing to the field of bio-inspired robotics.

Laws of Robotics

Laws of Robotics are a set of laws, rules, or principles, which are intended as a fundamental framework to underpin the behavior of robots designed to have a degree of autonomy. The best known set of laws is Isaac Asimov's "Three Laws of Robotics" (often shortened to The Three Laws or Asimov's Laws) in 1942. The Three Laws are described as below.

First: a robot may not injure a human being or, through inaction, allow a human being to come to harm.

Second: a robot must obey the orders given it by human beings except where such orders would conflict with the First Law.

Third: a robot must protect its own existence as long as such protection does not conflict with the First or Second Laws.

In later fiction, Asimov added a zeroth law, to precede the others:

A robot may not injure humanity, or, by inaction, allow humanity to come to harm.

The Laws have been used by many others to define laws used in fact and fiction.

Vocabulary 词汇

robot ['rəʊbɒt]	n. 机器人
machine [mə'ʃi:n]	n. 机器 v. 用机器制造
labor ['leɪbə]	n. 劳动；工作 vi. 劳动；努力
factory ['fæktəri]	n. 工厂；制造厂；代理店
describe [dɪ'skraɪb]	v. 描述，描写
efficient [ɪ'fɪʃənt]	adj. 有效率的；有能力的
emotionless [ɪ'məʊʃənlɪs]	adj. 没有情感的；不露情感的
incapable [ɪn'keɪpəbəl]	adj. 不能的；不能胜任的
engineering [ˌendʒɪ'nɪərɪŋ]	n. 工程；工程学
mechanics [mɪ'kænɪks]	n. 力学(用作单数)；结构；机械学(用作单数)
behavior [bɪ'heɪvjə]	n. 行为，举止；态度；反应
cognition [kɒg'nɪʃən]	n. 认识；知识；认识能力
autonomy [ɔ:'tɒnəmi]	n. 自治，自治权
injure ['ɪndʒə]	vt. 伤害，损害
inaction [ɪn'ækʃən]	n. 不活动；迟钝
obey [əʊ'beɪ]	vt. 服从，听从；按照……行动
protect [prə'tekt]	vt. 保护，防卫；警戒

humanity [hjuːˈmænəti]　　　*n.* 人类；人道；仁慈；人文学科
software [ˈsɒftweə]　　　　*n.* 软件
fiction [ˈfɪkʃən]　　　　　*n.* 小说；虚构；编造；谎言

Notes 注释

1. industrial robot　工业机器人
2. remote control　遥控；遥控装置；远程控制
3. do work by themselves　自主作业
4. bio-inspired robotics　仿生机器人
5. three laws of robotics　机器人三原则；机器人三定律

Part 2　Types and Application of Robot

Robotics is a rapidly growing field with the continuous advancement of technology. Now, robots can be found in homes as toys, vacuums, and programmable pets. Also, robots are an important part of many aspects of industry, medicine, science, space exploration, construction, food packaging and are even used to perform surgery.

There are many types of robots, such as mobile robots, industrial robots (manipulating), service robots, educational robots, modular robots, collaborative robots, etc, as shown in Figure 1-2 and Figure 1-3. They are used in many different environments and for many different purposes. Robots have replaced humans in performing repetitive and dangerous tasks, which humans prefer not to do, or are unable to do because of size limitation, or which take place in extreme environments such as outer space or the bottom of the sea.

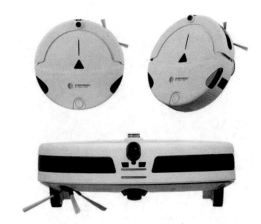

Figure 1-2　Articulated Welding Robots　　　Figure 1-3　Sweeping Robots

All in all, robots can be used in two categories of job.

(1) Jobs which a robot can do better than a human. Here, robots can increase productivity, accuracy, and endurance.

(2) Jobs which a human can do better than a robot, but it is desirable to remove the human for some reason. Here, robots free us from dirty, dangerous and dull tasks.

Vocabulary 词汇

vacuum [ˈvækjʊəm]　　　　　　　n. 真空；空间 adj. 真空的

repetitive [rɪˈpetətɪv]　　　　　　adj. 重复的

limitation [ˌlɪmɪˈteɪʃən]　　　　　n. 限制；限度

productivity [ˌprɒdʌkˈtɪvəti]　　　n. 生产力；生产率；生产能力

endurance [ɪnˈdjʊərəns]　　　　　n. 忍耐力；忍耐；持久；耐久

dirty [ˈdəːti]　　　　　　　　　　　　adj. 卑鄙的；肮脏的 vt. 弄脏 vi. 变脏
industry [ˈɪndəstri]　　　　　　　　　n. 产业；工业；勤勉
medicine [ˈmedsən]　　　　　　　　　n. 药；医学 vt. 用药物治疗
science [ˈsaɪəns]　　　　　　　　　　n. 科学；技术；学科
construction [kənˈstrʌkʃən]　　　　n. 建设；建筑物；解释；造句
perform [pəˈfɔːm]　　　　　　　　　 vt. 执行；完成 vi. 机器运转
surgery [ˈsəːdʒəri]　　　　　　　　　n. 外科；外科手术；手术室
replace [rɪˈpleɪs]　　　　　　　　　　v. 取代；替换

Notes 注释

1. programmable pet　可编程宠物
2. space exploration　太空探索
3. food packaging　食品包装
4. mobile robot　移动机器人
5. service robot　服务用机器人
6. educational robot　教育机器人
7. modular robot　模块化机器人
8. collaborative robot　协作机器人
9. size limitation　尺寸限制
10. extreme environment　极端环境
11. outer space　外太空；外层空间
12. all in all　总而言之

Part 3 About Industrial Robot

Definition and Application

Industrial robot is a robot system used for manufacturing. Industrial robot is defined as "an automatically controlled, reprogrammable, multipurpose, manipulator programmable in three or more axes, which may be either fixed in place or mobile for use in industrial automation applications." in ISO 8373.

The Robotics Institute of America(RIA) defines a robot as: A reprogrammable multi-functional manipulator designed to move materials, parts, tools, or specialized devices through variable programmed motions for the performance of a variety of tasks.

Industrial robot is helpful in material handling and providing interfaces. Typical applications of robots include welding, painting, assembly, pick and place for PCB, packaging, labeling, palletizing, inspection and testing; all can be accomplished with high endurance, speed, and precision.

History of Industrial Robot

In 1959, George Devol invented Unimate, the first industrial robot, as shown in Figure 1-4. It was used in a production line in the General Motors Corporation in 1961. It was used to lift red-hot door handles and other car parts from die-casting machines and stack them, in a factory in New Jersey, USA. So the first Unimate was a material handling robot. Its most distinctive feature was a gripper that eliminated the need for man to touch car parts just made from molten metal.

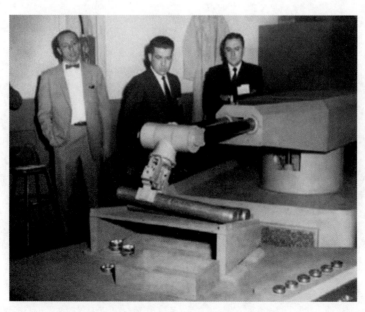

Figure 1-4 The First Industrial Robot, Unimate

The robot was soon followed by robots for welding and other applications, which undertook the job of transporting die-castings from an assembly line and welding these parts on auto bodies, a dangerous task for workers, who might be poisoned by toxic fumes or lose a limb if they were not careful.

Commercial Industrial Robots

In 1962, AMF Corporation produced the Verstran robot, which became a commercial industrial robot, as well as Unimate, and exported to countries around the world, setting off a robot boom worldwide.

Around 1970s, many companies started their robotic business and created their first industrial robot. Such as Nachi, KUKA, FANUC, YASKAWA, ASEA(predecessor of ABB) and OTC.

In 1978, Unimation created the PUMA (programmable universal machine for assembly) robot with support from General Motors, as shown in Figure 1-5. PUMA is still working in the production, many researches of industrial robots are based on the robot's model and object.

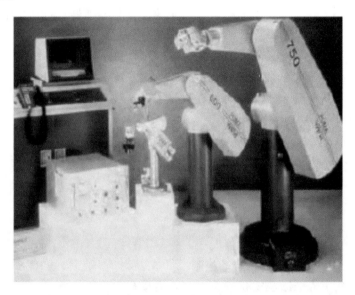

Figure 1-5　PUMA Robot

In 1979, OTC was originally a supplier of welding equipment, it expanded to become a provider to the Japanese auto market of GMAW supplies. Later, OTC Japan introduced its first generation of dedicated arc welding robots.

In 1980, The industrial robot industry started its rapid growth, with a new robot or company entering the market every month.

Vocabulary 词汇

manufacturing [ˌmænjʊˈfæktʃərɪŋ]	n. 制造业;工业 v. 制造;生产
automatic [ˌɔːtəˈmætɪk]	adj. 自动的;无意识的;必然的
reprogrammable	adj. 可改编程序的;可重复编程的
materials [məˈtɪəriəlz]	n. 材料;材料科学;材料费

interface [ˈɪntəfeɪs]　　　　　　　　　　n. 接口 v. 使联系
typical [ˈtɪpɪkəl]　　　　　　　　　　　adj. 典型的；特有的；象征性的
labeling [ˈleɪblɪŋ]　　　　　　　　　　n. 标签；标记 v. 贴标签；分类
inspection [ɪnˈspekʃən]　　　　　　　　n. 视察；检查
accomplished [əˈkʌmplɪʃt]　　　　　　 adj. 完成的；有技巧的
distinctive [dɪˈstɪŋktɪv]　　　　　　 adj. 有特色的，与众不同的
poison [ˈpɔɪzən]　　　　　　　　　　　n. 毒药；有毒害的事物
commercial [kəˈmɜːʃəl]　　　　　　　　adj. 商业的；营利的 n. 商业广告
Nachi　　　　　　　　　　　　　　　　n. 不二越（日本轴承品牌）
KUKA　　　　　　　　　　　　　　　　 n. 库卡（德国公司，后被美的收购）
FANUC　　　　　　　　　　　　　　　　n. 发那科（日本公司）
YASKAWA　　　　　　　　　　　　　　　n. 日本安川电气
predecessor [ˈpriːdɪsesə]　　　　　　 n. 前任，前辈
red-hot [redˈhɒt]　　　　　　　　　　 adj. 炽热的；最新的
gripper [ˈɡrɪpə]　　　　　　　　　　　n. 夹子，钳子；抓器，抓爪

Notes 注释

1. robot system　机器人系统
2. manipulator programmable　可编程机械手
3. The Robotics Institute of America　美国机器人研究所
4. production line　生产线
5. die-casting machine　压铸机
6. molten metal　熔融金属
7. assembly line　流水线
8. toxic fume　有毒烟雾
9. Verstran robot　Verstran 机器人
10. automatically controlled　自动控制
11. industrial automation application　工业自动化应用
12. robotics institute　机器人协会
13. material handling robot　材料处理机器人
14. assembly line　装配线
15. robot boom　机器人热潮
16. robotic business　机器人业务
17. robot model　机器人模型
18. arc welding robot　电弧焊机器人

Part 4 Component and Applications of Industrial Robot

The most commonly used industrial robots are articulated robots, SCARA robots, Delta robots and Cartesian coordinate robots (gantry robots or X-Y-Z robots).

Component

A typical industrial robot consists of a robotic arm, a control system, a teach pendant, an end effector as well as some other peripheral devices, as shown in Figure 1-6.

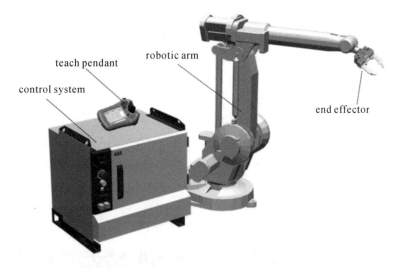

Figure 1-6 Components of Industrial Robot

The robotic arm, basically, is the part that moves the tool, but not every industrial robotic arm resembles an arm, there are different types of different robot structures. The control system resembles robot's brains, the teach pendant makes up the user environment. The teach pendant is usually used only in time of programming. The end effector is a device designed for specific tasks, for example, welding and painting.

Types

The main types of industrial robots are described as below.

1. Cartesian Robot/Gantry Robot

Used for pick and place work, application of sealant, assembly operations and welding. It's a robot whose arm has three prismatic joints, whose axes are coincident with a Cartesian coordinate.

2. Cylindrical Robot

Used for assembly operations, handling machine tools, spot welding, and handling die-casting machines. It's a robot whose axes form a cylindrical coordinate system.

3. Spherical Robot

Used for handling machine tools, spot welding, die-casting, fettling machines, gas welding and arc welding. It's a robot whose axes form a polar coordinate system.

4. SCARA Robot

Used for pick and place work, application of sealant, assembly operations and handling machine tools. This robot has two parallel rotary joints.

5. Articulated Robot

Used for assembly operations, die-casting, fettling machines, welding and spray painting. It's a robot whose arm has at least three rotary joints.

6. Parallel Robot

It's a robot whose arms have concurrent prismatic or rotary joints.

Application

Here are the top 4 applications for industrial robots.

1. Robotic Handling

As shown in Figure 1-7, material handling is the most popular application of industrial robots worldwide. This includes robotic machine tending, various operations of metal machining and plastic moulding. With the introduction of collaborative robots in the last few years, the market of robotic handling is continuously increasing.

2. Robotic Welding

As shown in Figure 1-8, this segment mostly includes spot welding and arc welding which is mainly used in the automotive industry. More and more small work shops are beginning to introduce welding robots into their production. In fact, with the price of robots going down and the various tools now available on the market, it is easier to automate a welding process.

Figure 1-7　Robotic Handling

Figure 1-8　Robotic Welding

3. Robotic Assembly

As shown in Figure 1-9, assembly operations include fixing, press-fitting, inserting, disassembling, etc. This category of robotic applications seems to have decreased over the last few years, even while other robotic applications have increased. The reason why the applications are diversified is the introduction of different technologies such as force torque sensors and tactile sensors that give more sensations to the robot.

4. Robotic Dispensing

As shown in Figure 1-10, here we are talking about painting, gluing, applying adhesive sealing, spraying, etc. Only 4% of the robots are doing dispensing. The fluency of robot makes a repeatable and accurate process.

Figure 1-9　Robotic Assembly　　　　　　　Figure 1-10　Robotic Dispensing

Vocabulary 词汇

brain [breɪn]	n. 头脑, 智力; 脑袋
die-casting [daɪˈkɑːstɪŋ]	压铸; 压铸件; 压铸作业
fettling [ˈfetlɪŋ]	n. 补炉材料; 铸件清理
axes [ˈæksiːz]	n. 轴线; 轴心; 坐标轴
fixing [ˈfɪksɪŋ]	n. 安装; 设备; 修理 v. 固定
inserting [ɪnˈsəːtɪŋ]	n. 插入; 嵌进 v. 插入
disassembling [ˌdɪsəˈsemblɪŋ]	n. 拆卸 v. 拆开 adj. 拆卸的
dispensing [dɪsˈpensɪŋ]	n. 配药; 调剂 v. 分发; 执行
gluing [ɡluːɪŋ]	n. 黏合; 胶合
apply [əˈplaɪ]	vt. 申请; 敷 vi. 申请; 敷; 适用

Notes 注释

1. press-fitting　压装; 配合
2. articulated robot　铰接式机器人, 关节型机器人
3. SCARA robot　SCARA 机器人
4. Delta robot　Delta 机器人
5. Cartesian coordinate robot　笛卡尔坐标机器人
6. gantry robot　龙门式机器人
7. control system　控制系统
8. teach pendant　示教盒
9. end effector　末端执行器
10. peripheral equipment　外围设备
11. robot structure　机器人结构

12. the user environment   用户环境
13. pick and place   挑选和放置；拾取和放置
14. prismatic joint   棱柱形接头
15. cylindrical robot   圆柱形机器人
16. cylindrical coordinate system   圆柱坐标系
17. spherical robot   球形机器人
18. gas welding   气焊
19. arc welding   电弧焊
20. polar coordinate system   极坐标系
21. two parallel rotary joints   两个平行的旋转关节
22. spray painting   喷漆
23. parallel robot   并联机器人
24. material handling   物料搬运
25. plastic moulding   塑料成型
26. spot welding   点焊
27. automotive industry   汽车行业
28. go down   下降
29. force torque sensor   力矩传感器
30. tactile sensor   触觉传感器
31. adhesive sealing   胶黏剂密封
32. fettling machine   保养机器人
33. robotic machine tending   机器人管理

Unit 2　Introduction of Industrial Robot

Part 1　Main Parts of Industrial Robot

Robots can be made from a variety of materials including metals and plastics. The industrial robot is composed of 4 parts: manipulator, controller, demonstrator and robotic hand, as shown in Figure 2-1.

Figure 2-1　Main Parts of Industrial Robot

(1) Manipulator is also known as the robotic arm, a type of mechanical arm, usually programmable, with similar functions to a human arm. The arm may be the sum total of the mechanism or may be part of a more complex robot. Robotic arm is connected by joints, which allow either rotational motion or translational(linear) displacement. The terminus of the manipulator is an end effector.

(2) Controller is also known as the "brain" which is run by a computer program. Usually, the program is very detailed as it gives commands for the moving parts of the robot to follow.

(3) Demonstrator is also known as the teaching box, a human-computer interaction interface, connected with the controller, can be operated to move by the operator.

(4) Robotic hand is the end effector, or end-of-arm-tooling (EoAT), can be designed to perform any desired tasks, such as welding, gripping, spinning etc, depending on the application. End effectors are generally highly complex, made to match the handled products and often capable

of picking up an array of products at one time. They may utilize various sensors to aid the robot system in locating, handling, and positioning products.

All of these parts work together to control how the robot operates.

Vocabulary 词汇

demonstrator [ˈdemənstreɪtə]	n. 示威者；论证者；指示者
programmable [ˈprəʊɡræməbəl]	adj. 可编程的；可设计的
joint [dʒɔɪnt]	n. 接头；关节；运动副
interaction [ˌɪntərˈækʃən]	n. 相互作用；交互作用
grip [ɡrɪp]	vt. 紧握；夹紧 vi. 抓住
spin [spɪn]	v. 旋转；纺织
utilize [ˈjuːtɪlaɪz]	vt. 利用
locating [ləʊˈkeɪtɪŋ]	n. 定位 v. 找出；安置
position [pəˈzɪʃən]	vt. 安置；把……放在适当位置

Notes 注释

1. industrial robot 工业机器人
2. be made from… 由……做成
3. a variety of 各种各样
4. robotic hand 机器手
5. rotational motion 旋转运动
6. translational(linear)displacement 平移(线性)位移
7. end effector 末端执行器
8. teaching box 示教盒
9. human-computer interaction interface 人机交互界面

Part 2 Basic Terms

1. Rigid Body

In physics, a rigid body is an idealization of a solid body whose deformation can be neglected. In other words, the distance between any two given points of a rigid body remains constant all the time regardless of external forces exerted on it.

2. Rotary Joint

A rotary joint (also called pin joint or hinge joint) is a one degree of freedom kinematic pair used in mechanisms. Rotary joints provide single-axis rotation function used in many places such as door hinges, folding mechanisms, and other uni-axial rotation devices, as shown in Figure 2-2.

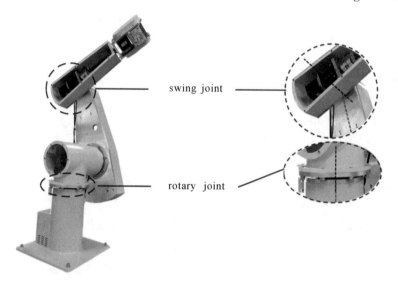

Figure 2-2 Rotary Joint of EDUBOT-PUMA 560

3. Kinematic Pair

A kinematic pair is a connection between two bodies that impose constraints on their relative movements.

4. Articulated Robot

Articulated robot is the one that uses rotary joints to access its work space. Usually the joints are arranged in a "chain", so that one joint supports another further in the chain.

5. Continuous Path

It is a control scheme by the inputs or commands that specifies every point moves along a desired path of motion. The path is controlled by the coordinated motion of the manipulator joints.

6. Kinematics

The actual arrangement of rigid members and joints in the robot determines the robot's possible motions. The kinds of robot kinematics include articulated, Cartesian, parallel and

SCARA.

7. Motion Control

For some applications, such as simple pick and place, assembly, the robot need merely return repeatedly to a limited number of pre-taught positions. For more sophisticated applications, such as welding and finishing(spray painting), motion must be continuously controlled to follow a path in space, with controlled orientation and velocity.

8. Power Source

Some robots use electric motors, others use hydraulic actuators. The former are faster, the latter are stronger and more advantageous in applications such as spray painting, where a spark could set off an explosion.

9. Drive

Some robots connect electric motors to the joints via gears; others connect the motors to the joints directly(direct drive). Using gears results in measurable "backlash" which is a free movement in an axis. Small robot arms frequently employ high speed, low torque DC motors, which generally require high gearing ratios; the disadvantage is back lash. In such cases the harmonic drive is often used.

Vocabulary 词汇

deformation [ˌdiːʃəˈmeɪʃən]　　　　　　n. 变形
neglected [nɪˈɡlektɪd]　　　　　　　　adj. 被忽视的 v. 忽视;疏忽
pin [pɪn]　　　　　　　　　　　　　n. 针;栓 vt. 钉住;压住
hinge [hɪndʒ]　　　　　　　　　　　n. 铰链 vt. 给……安装铰链
chain [tʃeɪn]　　　　　　　　　　　　n. 链;束缚 vt. 束缚;囚禁
articulated [ɑːˈtɪkjʊleɪtɪd]　　　　　　adj. 铰接式的 v. 铰接
Cartesian [kɑːˈtɪzɪən]　　　　　　　　adj. 笛卡尔的 n. 笛卡尔信徒
finishing [ˈfɪnɪʃɪŋ]　　　　　　　　　adj. 最后的 n. 完成 v. 完成
backlash [ˈbækˌlæʃ]　　　　　　　　n. 反冲;强烈抵制 vt. 强烈反对
slightly [ˈslaɪtli]　　　　　　　　　　adv. 些微地,轻微地;纤细地
axis [ˈæksɪs]　　　　　　　　　　　n. 轴;轴线;轴心国 [复数]axes

Notes 注释

1. rigid body　刚体
2. external force　外力
3. rotary joint　旋转接头,旋转关节
4. pin joint　针接头
5. hinge joint　铰链接头
6. kinematic pair　运动对偶,运动副
7. relative movement　相对运动
8. articulated robot　铰接式机器人
9. continuous path　连续路径

10. motion control   运动控制
11. direct drive   直接驱动
12. given point   给定点
13. folding mechanism   折叠机构
14. uni-axial rotation   单轴旋转
15. control scheme   控制方案
16. hydraulic actuator   液压执行机构
17. harmonic drive   谐波驱动；谐波传动

Part 3 Technical Parameters

The technical parameters reflect the scope and performance of the robot, including number of axes, degrees of freedom, payload, working envelope, maximum speed, resolution, accuracy and repeatability. Other parameters are control mode, drive mode, installation, power source capacity, body weight and environmental parameters.

1. Number of Axes

Two axes are required to reach any point in a plane; three axes are required to reach any point in space. To fully control the orientation of the end of the arm (i. e. the wrist) three more axes are required. Some designs (eg the SCARA robot) trade limitations in motion possibilities for cost, speed, and accuracy.

2. Degrees of Freedom(DOF)

It is the number of independent motions that the end effector of the robot can move. It is defined by the number of axes of motion of the manipulator, usually the same as the number of axes, as shown in Figure 2-3. DOF reflects the flexibility of robot action. With more DOF, the robot is closer to human action and the versatility of the robot is better. However, the mechanisms are more complex and the overall requirements of the robot are higher. Therefore, the degrees of freedom of industrial robots is designed according to application.

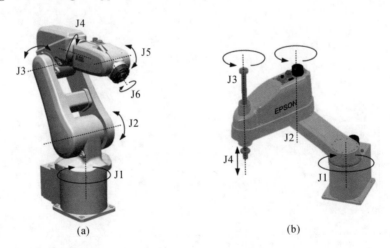

Figure 2-3   DOF of robot
(a) six-axis Robot   (b) EPSON SCARA Robot

3. Payload

It is the weight that a robot can lift. The maximum payload is the weight carried by the robot manipulator at reduced speed while maintaining rated precision. Nominal payload is measured at maximum speed while maintaining rated precision. The payload of different industrial robots is described in Table 2-1.

Table 2-1 Payload of Industrial Robots

Product Number	Payload	Image	Product Number	Payload	Image
ABB YuMi	0.5 kg		YASKAMA MH12	12 kg	
ABB IRB120	3 kg		YASKAWA MC2000ll	50 kg	
KUKA KR6	6 kg		FANUC R-2000iB/210F	210 kg	
KUKA KR16	16 kg		FANUC M-200iA/2300	2300 kg	

4. Working Envelope

It is also known as working space (reach envelope). A three-dimensional shape that defines the region of space a robot can reach, as shown in Figure 2-4.

5. Maximum Speed

It is the compounded maximum speed of the tip of a robot moving at full extension with all joints moving simultaneously in complimentary directions. This speed is the theoretical maximum and should not be used to estimate cycle time for a particular application.

6. Resolution

It is the smallest increment of motion or distance that can be detected or controlled by the control system of a mechanism. The resolution of any joint is a function of encoder pulses per revolution and drive ratio, and dependent on the distance between the tool center point and the joint axis.

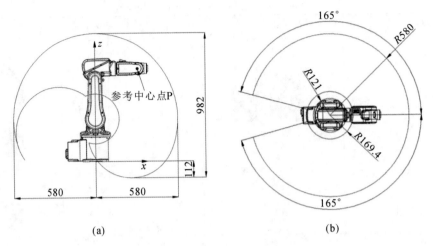

Figure 2-4　Working Envelope of ABBIRB 120
(a) Main View　(b) Top View

7. Accuracy

It is the closeness a robot can reach a commanded position. Accuracy is the difference between the point that a robot is trying to achieve and the actual resultant position. Absolute accuracy is the difference between a point instructed by the robot control system and the point actually achieved by the manipulator arm, while repeatability is the cycle-to-cycle variation of the manipulator arm when aimed at the same point.

8. Repeatability

It is the degree that the robot will return to a programmed position. That is to say, the ability of a system to repeat the same motion or achieve the same points when presented with the same control signals. It is the cycle-to-cycle error of a system when trying to perform a specific task.

Vocabulary 词汇

orientation [ˌɔːriən'teɪʃən]　　　　n. 方向；定向；情况介绍
flexibility [ˌfleksɪ'bɪlɪti]　　　　n. 灵活性；弹性；适应性
versatility [ˌvɜːsə'tɪləti]　　　　n. 多功能性；用途广泛
scope [skəʊp]　　　　n. 范围；余地；视野
payload ['peɪləʊd]　　　　n. 有效载荷
resolution [ˌrezə'luːʃən]　　　　n. 分辨率
repeatability [rɪˌpiːtə'bɪlɪti]　　　　n. 重复性；再现性
installation [ˌɪnstə'leɪʃən]　　　　n. 安装；装置；就职
increment ['ɪŋkrɪmənt]　　　　n. 增量；增加；增额

Notes 注释

1. technical parameter　技术参数
2. degrees of freedom　自由度
3. maximum payload　最大有效负载

4. nominal payload　额定负载
5. working envelope　工作空间
6. maximum speed　最大速度
7. environmental parameter　环境参数
8. drive ratio　驱动比
9. absolute accuracy　绝对精度

Part 4　Kinematics and Dynamics

The study of motion can be divided into kinematics and dynamics.

Robot Kinematics

Robot kinematics applies geometry to the study of the movement of multi degrees of freedom kinematic chains that form the structure of robotic systems. The emphasis on geometry means that the links of a robot are modeled as rigid bodies and its joints are assumed to provide pure rotation or translation.

Robot kinematics studies the relationship between the dimensions, connectivity of kinematic chains and the position, velocity, acceleration of each of the links in the robotic system, in order to plan and control movement and to compute actuator forces and torques.

A fundamental tool in robot kinematics is the kinematics equations of the kinematic chains that form the robot.

Forward kinematics uses the kinematic equations of a robot to compute the position of the end effector from specified values for the joint parameters, as shown in Figure 2-5. The reverse process that computes the joint parameters that achieve a specified position of the end effector is known as inverse kinematics, as shown in Figure 2-6. The dimensions of the robot and its kinematics equations define the volume of space reachable by the robot, known as its workspace.

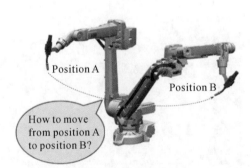

Figure 2-5　Forward Kinematics　　　　　　Figure 2-6　Inverse Kinematics

Forward kinematics specifies the joint parameters and computes the configuration of the chain. For serial manipulators this is achieved by direct substitution of the joint parameters into the forward kinematics equations for the serial chain. For parallel manipulators, the substitution of the joint parameters into the kinematics equations requires the solution of a set of polynomials constraints to determine the set of possible end effector locations.

Inverse kinematics specifies the end effector locations and computes the associated joint angles. For serial manipulators this requires the solution to a set of polynomials obtained from the kinematics equations and yields multiple configurations for the chain. For parallel manipulators, the specification of the end effector location simplifies the kinematics equations, which yields formulas for the joint parameters.

Forward kinematics refers to the calculation of end effector position, orientation, velocity, and acceleration when the corresponding joint values are known. Inverse kinematics refers to the opposite case in which required joint values are calculated for given end effector values, as done in path planning.

Forward dynamics refers to the calculation of accelerations in the robot once the applied forces are known. Forward dynamics is used in computer simulations of the robot.

Inverse dynamics refers to the calculation of the actuator forces necessary to create a prescribed end effector acceleration. This information can be used to improve the control algorithms of a robot.

Vocabulary 词汇

kinematics [ˌkɪnɪ'mætɪks]	n. 运动学；动力学
dynamics [daɪ'næmɪks]	n. 动力学；力学
motion ['məʊʃən]	n. 动作；移动；手势
geometry [dʒi'ɒmɪtri]	n. 几何学；几何结构
translation [træns'leɪʃən]	n. 翻译；译文；转化；调任
dimension [daɪ'menʃən]	n. 规模，大小
connectivity [kəˌnektɪvə'ti]	n. 连通性
velocity [vɪ'lɒsəti]	n. 速度
torque [tɔːk]	n. 扭转；转（力）矩
polynomial [ˌpɒlɪ'nəʊmɪəl]	n. 多项式
algorithm ['ælgərɪðəm]	n. 算法

Notes 注释

1. multi degrees of freedom   多自由度
2. kinematic chain   运动链
3. rigid body   刚体
4. kinematics equation   运动学方程
5. forward kinematics   正向运动学
6. inverse kinematics   反向运动学
7. serial manipulator   串联机械臂
8. parallel manipulator   并联机械臂
9. the corresponding joint   对应关节
10. path planning   路径规划
11. forward dynamics   正向动力学

12. inverse dynamics 反向动力学
13. the control algorithm 控制算法
14. serial chain 串行链
15. joint angle 关节角度

Unit 3   Types of Industrial Robots

Part 1   Cartesian Coordinate Robot

Cartesian coordinate robot(also called linear robot) is an industrial robot with simple structure, as shown in Figure 3-1.

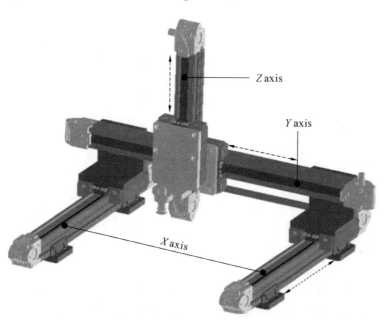

Figure 3-1   Cartesian Coordinate Robot

Cartesian coordinate robot is a robot whose arm has three prismatic joints. Its axes are coincident with a Cartesian coordinator; its three principal axes of control are linear (i. e. they move in a straight line rather than rotate) and are at right angles to each other. The three sliding joints correspond to moving the wrist up-down, in-out, back-forth. Among other advantages, this mechanical arrangement simplifies the robot control solution. Cartesian coordinate robots with horizontal members supported at both ends are sometimes called gantry robots. In terms of mechanical structure, they resemble gantry cranes, although the latter are not generally robots. Gantry robots are generally quite large.

A popular application for Cartesian coordinate robot is a computer numerical control machine (CNC machine) and 3D printing, as shown in Figure 3-2. The simplest application is used in milling and drawing machines where a pen or a router translates across an X-Y plane while a tool is raised and lowered onto a surface to create a precise design. Pick and place machines and plotters are also based on the principle of Cartesian coordinate robot, as shown in Figure 3-3.

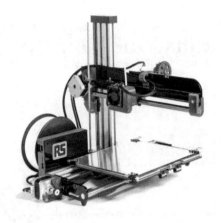

Figure 3-2　3D Printing

Figure 3-3　Pick and Place Machine

Vocabulary 词汇

prismatic [prɪzˈmætɪk]　　　　　*adj.* 棱镜的；光彩夺目的
structure [ˈstrʌktʃə]　　　　　　*n.* 构造；建筑物
milling [ˈmɪlɪŋ]　　　　　　　　*n.* 磨；制粉；轧齿边 *v.* 碾磨
drawing [ˈdrɔːɪŋ]　　　　　　　*n.* 图画；牵引 *v.* 绘画；吸引
router [ˈruːtə]　　　　　　　　 *n.* [计]路由器；剜刨工具
plotter [ˈplɒtə]　　　　　　　　*n.* [测]绘图仪；阴谋者

Notes 注释

1. linear robot　线性机器人
2. rather than　而不是
3. sliding joint　滑动接头
4. gantry robot　龙门式机器人
5. computer numerical control machine　计算机数控机床
6. mechanical structure　机械结构
7. drawing machine　拉丝机
8. pick and place machine　拾放机

Part 2　SCARA Robot

The SCARA acronym stands for selective compliance assembly robot arm or selective compliance articulated robot arm.

In 1981, SCARA (as shown in Figure 3-4) was developed under the guidance of Hiroshi Makino, a professor at the University of Yamanashi.

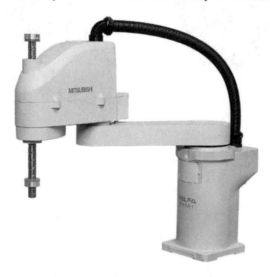

Figure 3-4　SCARA Robot

Attributes

Its arm is slightly compliant in the XY-axes but rigid in the Z-axis. This feature is particularly suitable for assembly work, for example, inserting a round pin into a round hole. Therefore, the SCARA system is widely used for assembling printed circuit boards and electronic components.

The other attribute of SCARA is the jointed two-link arm layout which is similar to human arms, hence it often uses the term "articulated". This feature allows the arm to extend into confined areas and then retract or "fold up" out of the way. This is advantageous for transferring parts from one cell to another or loading/unloading process stations that are enclosed.

Features

SCARAs are generally faster and cleaner than comparable Cartesian robot systems. Their single pedestal mount requires a small footprint and provides an easy, unhindered form of mounting. But SCARAs can be more expensive than comparable Cartesian systems and the controlling software requires inverse kinematics for linear interpolated moves. The software usually comes with SCARA and is transparent to the end-user.

Most SCARA robots are based on serial architectures, which means that the first motor should carry all other motors. There also exists a so-called double-arm SCARA robot architecture, in which two of the motors are fixed on the base. The first such robot was commercialized by Mitsubishi Electric; another example of a dual-arm SCARA robot is Mecademic's DexTAR educational robot. In China, HRG company has developed five-bar robot HRG-HD1B5B, as shown in Figure 3-5.

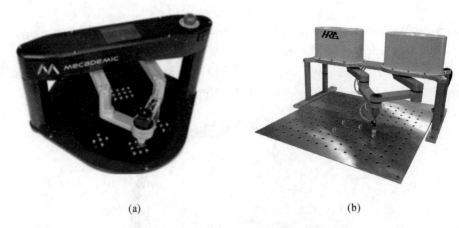

Figure 3-5　Dual-arm SCARA robot
(a) DexTAR Educational Robot　(b) HRG-HD1B5B Five-bar Robot

Vocabulary 词汇

selective [sɪ'lektɪv]　　　　adj. 选择性的；讲究的
attribute [ə'trɪbjuːt]　　　　n. 属性
compliant [kəm'plaɪənt]　　adj. 顺从的；服从的；应允的
enclose [ɪn'kləʊz]　　　　vt. 把……装入信封；装入
feature ['fiːtʃə]　　　　　　n. 产品特点，特征 v. 是……的特色，使突出
mount [maʊnt]　　　　　　vt. 增加；安装，架置 n. 山峰；底座
footprint ['fʊtprɪnt]　　　　n. 足迹，脚印
end-user ['end'juːzə]　　　　n. 终端用户
rigid ['rɪdʒɪd]　　　　　　　adj. 僵硬的；坚硬的；精确的

Notes 注释

1. serial architecture　串行架构
2. double-arm SCARA robot　双臂 SCARA 机器人
3. DexTAR educational robot　DexTAR 教育机器人
4. articulated robot　关节型机器人
5. assembly robot　装配机器人
6. printed circuit board　印刷电路板
7. electronic component　电子零部件
8. two-link arm　联合双链臂
9. process station　处理站

Part 3 Six-axis Articulated Robot

Industrial robots have various axis configurations. The articulated robot is a kind of robot (eg legged robot or industrial robot) with rotary joints. Articulated robots can range from simple two-jointed structures to systems with 10 or more interacting joints. They are powered by a variety of means, including electric motors.

The six-axis articulated robot is a kind of widely used mechanical equipment. The vast majority of articulated robots have six axes, also called six degrees of freedom, as shown in Figure 3-6. Six-axis robots allow for greater flexibility and can perform a wider variety of applications than robots with fewer axes.

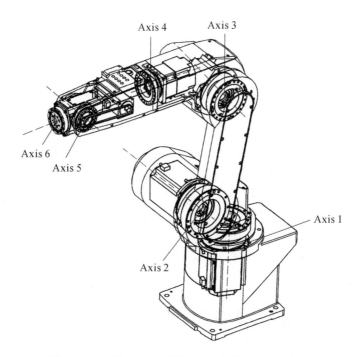

Figure 3-6 Six Axes of Six-axis Articulated Robot

Axis 1—This axis is located on the robot base and allows the robot to rotate from left to right. This sweeping motion extends the work area to include the area on either side and behind the arm. This axis allows the robot to spin up to a full 180 degree range from the center point. This axis is also known as J1.

Axis 2—This axis allows the lower arm of the robot to extend forward and backward. It is the axis powering the movement of the entire lower arm. This axis is also known as J2.

Axis 3—This axis extends the robot's vertical reach. It allows the upper arm to raise and descend. In some articulated models, it allows the upper arm to reach behind the body, further expanding the work scope. This axis gives the upper arm the better part access.

This axis is also known as J3.

Axis 4—This axis works in conjunction with axis 5; it aids in the positioning of the end effector and manipulation of the part. Known as the wrist roll, it rotates the upper arm in a circular motion moving parts between horizontal to vertical orientations. This axis is also known as J4.

Axis 5—This axis allows the wrist of the robot arm to tilt up and down. This axis is responsible for the pitch and yaw motion. The pitch, or bend, motion is up and down, much like opening and closing a box lid. Yaw moves left and right, like a door with hinges. This axis is also known as J5.

Axis 6—This is the wrist of the robot arm. It is responsible for twisting motion, allowing it to rotate freely in a circular motion, both to position end effectors and to manipulate parts. It is usually capable of more than a 360 degree rotation in either a clockwise or counter clockwise direction. This axis is also known as J6.

Vocabulary 词汇

legged [legɪd]　　　　　　　　　　　　adj. 有腿的 v. 用脚移动或推动
interacting [ˌɪntərˈæktɪŋ]　　　　　　n. 相互作用 v. 互相影响
vertical [ˈvɜːtɪkəl]　　　　　　　　　adj. 垂直的;[解剖]头顶的 n. 垂直面
tilt [tɪlt]　　　　　　　　　　　　　 vi. 倾斜;翘起 vt. 使倾斜;使翘起 n. 倾斜
position [pəˈzɪʃən]　　　　　　　　　n. 定位;配置;布置 v. 定位;放置
manipulation [məˌnɪpjʊˈleɪʃnɛ]　　　 n. 操纵;操作;处理;篡改
commercialized [kəˈmɜːʃəlaɪzd]　　　adj. 商业化的 v. 使商品化
pitch [pɪtʃ]　　　　　　　　　　　　 vi. 倾斜;投掷 vt. 投;掷 n. 投掷

Notes 注释

1. legged robot　腿式机器人
2. robot base　机器人底座
3. lower arm　下臂
4. upper arm　上臂
5. six degrees of freedom　六自由度
6. articulated model　关节模型
7. dual-arm　双臂
8. yaw motion　偏航运动
9. bend motion　弯曲运动
10. manipulate part　操纵零件

Part 4 Palletizing Robot

Industrial palletizing refers to loading and unloading parts, boxes or other items to or from pallets. Automated palletizing refers to an industrial robot palletizer performing the application automatically.

Palletizing robots can be seen in many industries including food processing, manufacturing, and shipping. There is a large variety of robotic palletizers available with a large range of payload and reach, as shown in Figure 3-7 and Figure 3-8.

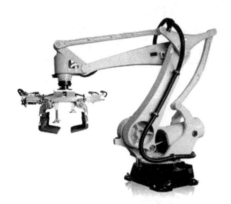

Figure 3-7 FANUC M-410iW   Figure 3-8 ABB IRB 260

Various end-of-arm-tooling styles allow flexibility of different types of robot palletization. Bag grippers encompass an item and support it on the bottom, while suction and magnetic grippers typically handle more ridged items and grip them from the top. By automating your shop with a palletizing robot, you can increase the consistency of your loading and unloading processes.

Palletizing robot was introduced in the early 1980s and has an end of arm tool (end effector) to grab the product from a conveyor or layer table and position it onto a pallet, as shown in Figure 3-9.

Palletizing robots provide speed, repeatability and precision at a low cost. With Palletizing robots, pallets are stacked and un-stacked in a systematic neat manner.

Palletizing robots systems have many advantages as below.

(1) Safety. Palletizing robot applications can protect workers. Robots are able to work in refrigerated and freezer environments. Palletizing robots can handle tasks that cause repetitive back injuries for workers. Robots lift heavy payloads and further speed up production.

(2) Flexibility. Palletizing robots make it easy to change a product or task. They are quickly reprogrammed to accommodate a wide variety of products and packaging. Interchangeable EOAT options allow robots to handle different materials or switch back

Figure 3-9 Palletizing Robot

and forth between applications. Advanced vision technology can allow robots to distinguish between parts and palletizing different items.

The palletizing robot will not only increase the speed of production line, but will also increase productivity.

Vocabulary 词汇

load [ləʊd]	vt. 装填;装满 vi. 加载;装载;装货
pallet ['pælɪt]	n. 货板;简陋小床;调色板
unload [ˌʌn'ləʊd]	vt. 卸货;倾吐;处理;倾销
shipping ['ʃɪpɪŋ]	n. 运送;[船]船舶;航运,海运
precision [prɪ'sɪʒən]	n. 精度 adj. 精密的;精确的
stacked [stækt]	adj. 妖艳的 v. 堆放
un-stacked	拆垛
refrigerated [rɪ'frɪdʒəreɪtɪd]	adj. 冷冻的,冷却的 v. 冷藏
switch [swɪtʃ]	vt. 转换 vi. 转换;抽打;换防 n. 开关;转换;鞭子
vision ['vɪʒən]	n. 视力;美景;眼力;幻象;想象力 vt. 想象;显现;梦见
technology [tek'nɒlədʒi]	n. 技术;工艺;术语
line [laɪn]	n. 路线;排;绳 vt. 排成一行 vi. 排队;站成一排

Notes 注释

1. automated palletizing　自动码垛
2. end of arm tool　端部执行器
3. vision technology　视觉技术

Part 5 Delta Robot

The Delta robot is a type of parallel robot. It consists of three arms connected to universal joints at the base. The key design feature is the use of parallelograms in the arms, which maintains the orientation of the end effector. By contrast, the Stewart platform can change the orientation of its end effector.

Typical Product

In 1999, ABB Flexible Automation started to sell its Delta robot, the FlexPicker, as shown in Figure 3-10.

In 2009, FANUC released the newest version of Delta robot, the FANUC M-1iA robot, and would later release variations of this Delta robot for heavier payloads. FANUC released the M-3iA in 2010 for heavier payloads, as shown in Figure 3-11. And most recently FANUC released the M-2iA Robot for medium-sized payloads in 2012.

Figure 3-10 ABB FlexPicker   Figure 3-11 FANUC M-3iA Robot

Design Principle

The Delta robot is a parallel robot. It consists of multiple kinematic chains connecting the base with the end effector. The robot can also be seen as a spatial generalisation of a four-bar linkage.

The key concept of the Delta robot is the use of parallelograms which restrict the movement of the end platform to pure translation, i.e. only to move in the X, Y or Z direction with no rotation.

The robot base is mounted above the workspace and all the actuators are located on it. From the base, three middle jointed arms extend. The end of these arms is connected to a small triangular platform. Actuation of the input links will move the triangular platform along the X, Y or Z direction. Actuation can be done with linear or rotational actuators, with or without reductions (direct drive).

Since the actuators are all located on the base, the arms can be made of a light composite material. As a result of this, the moving parts of the Delta robot have a small inertia. This allows for very high speed and high acceleration. Having all the arms connected together to the end effector increases the robot stiffness, but reduces its working volume.

Applications

Delta robots have popular usage in picking and packaging in factories because they can be quite fast, some can execute up to 300 picks per minute.

Industries that take advantage of the high speed of Delta robots are the packaging industry, medical and pharmaceutical industry. Because of its stiffness it is also used for surgery. Other applications include high precision assembly operations in a clean room for electronic components.

Vocabulary 词汇

parallelogram [ˌpærəˈleləˌɡræm]	n. 平行四边形
restrict [rɪˈstrɪkt]	vt. 限制；约束；限定
rotation [rəʊˈteɪʃn]	n. 旋转；循环，轮流
triangular [traɪˈæŋɡjulə]	adj. 三角的；[数]三角形的；三人间的
reduction [rɪˈdʌkʃən]	n. 减少；下降；还原反应
inertia [ɪˈnɜːʃə]	n. [力]惯性；迟钝；不活动
stiffness [ˈstɪfnɪs]	n. 僵硬；坚硬；不自然；顽固
workspace [ˈwɜːkspeɪs]	n. 工作空间

Notes 注释

1. Stewart platform   Stewart 平台
2. parallel robot   并联机器人
3. spatial generalization   空间泛化
4. four-bar linkage   四杆联动
5. composite material   复合材料
6. medical and pharmaceutical industry   医药行业
7. clean room   洁净室；无尘室
8. working volume   工作量

Unit 4　ABB Robot

Part 1　ABB and ABB Robot

ABB

ABB (ASEA Brown Boveri) is a Swiss multinational corporation headquartered in Zurich, Switzerland, operating mainly in robotics and the power and automation technology areas. ABB resulted from the merger of the Swedish corporation Allmä-nna Svenska Elektriska Aktiebolaget(ASEA) and the Swiss company Brown, Boveri & Cie(BBC) in 1988, as shown in Figure 4-1.

Figure 4-1　ABB

ABB is one of the largest engineering companies, its core businesses is power and automation technologies.

ABB Robot

ABB is a world leading manufacturer of industrial robots and robot systems, operating in 53 countries, in over 100 locations around the world.

In addition to robots, ABB manufactures and supplies robot software, peripheral equipment, process equipment, modular manufacturing cells and service for tasks such as welding, material handling, small parts assembly, painting & finishing, picking, packing, palletizing and machine tending.

The key markets of ABB robots include automotive, plastic, metal fabrication, foundry, solar, consumer electronics, wood, machine tool, pharmaceutical, food and beverage industries. A strong solution helps manufacturers improve productivity, product quality and worker safety. ABB has installed more than 250 000 robots worldwide.

ABB robots play a significant role in our daily lives; hardly a moment goes by without our using a product that was manufactured or handled by ABB robots. For instance, ABB robots pick, pack and palletize food and beverages for companies like Nestlé, Unilever and Cadbury. They carve, sand, finish, paint and package the furniture and flooring for two of the biggest companies in the business—Ikea and Tarkett; and they weld, grind, polish and paint PCs, laptops, iPods, mobile phones, cameras and game consoles for the world's leading brands and manufacturers—Apple, Dell, Nokia and many more, as shown in Figure 4-2.

Figure 4-2　Industrial Applications of Robots

In fact, ABB robots not only boost industrial productivity, they can also achieve a massive improvement in energy efficiency and greenhouse gas emission reduction. The FlexPainter IRB 5500 paint robot, for instance, has reduced paint shop energy consumption by 50 percent at automotive factories all over the world.

Vocabulary 词汇

headquartered [ˈhedˌkwɔːtəd]	adj. 以……为总部所在地的
merger [ˈmɜːdʒə]	n. (企业等的)合并;并购
peripheral [pəˈrɪfərəl]	adj. 外围的;次要的;(神经)末梢区域的 n. 外部设备
fabrication [ˌfæbrɪˈkeɪʃən]	n. 制造,建造;装配;伪造物
foundry [ˈfaʊndri]	n. 铸造,铸造类;铸造厂
robotics [rəʊˈbɒtɪks]	n. 机器人学
wood [wʊd]	n. 木材 vi. 收集木材 vt. 植林于
pharmaceutical [ˌfɑːməˈsjuːtɪkəl]	adj. 制药(学)的 n. 药物
beverage [ˈbevərɪdʒ]	n. 饮料
palletize [ˈpælɪˌtaɪz]	vt. 用托盘装;码垛堆积
carve [kɑːv]	vt. 雕刻;切开;开创 vi. 切开;做雕刻工作
finish [ˈfɪnɪʃ]	vt. 完成;结束 vi. 结束 n. 结束
boost [buːst]	vt. 促进;增加 vi. 宣扬;偷窃 n. 推动;帮助;宣扬
Cadbury [ˈkædbəri]	n. 吉百利
weld [weld]	vt. 焊接 vi. 焊牢 n. 焊接

Notes 注释

1. automation technology   自动化技术
2. process equipment   工艺设备
3. modular manufacturing cell   模块化制造单元
4. machine tending   机器管理
5. small parts assembly   小零件组装
6. play a significant role in…   扮演……重要角色;发挥重要作用
7. for instance   例如
8. energy efficiency   能效
9. greenhouse gas   温室气体
10. energy consumption   能耗

Part 2　ABB Product Series

Since 1988, in order to meet the market demand, ABB developed a series of industrial robot products, such as handling robots, welding robots, assembly robots, painting robots and so on. The following is a brief description of the main models of ABB robots (the specific parameters of the specifications are from ABB officially announced data).

IRB 1410

The IRB 1410 is optimized and designed for arc welding. It can handle a payload of 5 kg with a reach of 1 440 mm. The IRB 1410 provides functional package of arc welding; it can be manipulated through the teaching pendant and is widely used in arc welding, material handling and process applications, as shown in Figure 4-3.

IRB 2400

The IRB 2400 in its different versions and best accuracy, gives excellent performance in material handling, machine tending and process applications. The IRB 2400 offers increased production rates, reduced lead time and faster delivery for manufactured products. The IRB 2400 is one of the most popular industrial robots in the world, and it maximizes the efficiency of applications such as arc welding, machining, loading and unloading, as shown in Figure 4-4.

IRB 52

The IRB 52 is a compact painting robot designed specifically for consumable parts painting in general industries. With its small size and large working envelope, it is flexible and versatile, while its high speed and accuracy offers short cycle time. It includes ABB's unique integrated paint system (IPS) to ensure high quality, high precision process adjustments, and ultimately high-quality coating and reduced coating consumption, as shown in Figure 4-5.

Figure 4-3　IRB 1410 Robot　　Figure 4-4　IRB 2400 Robot　　Figure 4-5　IRB 52 Robot

IRB 360

For nearly 15 years, ABB's IRB 360 FlexPicker has been the leader in state-of-the-art high speed robotic picking and packing technology. Compared to conventional hard automation, the IRB 360 offers much greater flexibility in a compact footprint while maintaining accuracy and high payloads.

The IRB 360 family now includes variants with payloads of 1 kg, 3 kg, 6 kg and 8 kg and reaches of 800 mm, 1 130 mm and 1 600 mm—IRB 360 can meet almost every need. Featuring outstanding motion control, short cycle time, and precision accuracy, the IRB 360 can operate at very high speed in anything from narrow to wide space with very tight tolerances. Every FlexPicker also benefits from a re-engineered tool flange which can accommodate larger grippers, allowing for efficient handling of flow wrapped products at high speed from an indexing belt, as shown in Figure 4-6.

IRB 910SC

In designing its selective compliance articulated robot arm(SCARA), or IRB 910SC, ABB has delivered a single arm robot capable of operating in a confined footprint. ABB's SCARA is ideal for the small parts assembly, material handling and parts inspection, as shown in Figure 4-7.

ABB's SCARA family is designed for a variety of general-purpose applications such as tray kitting, component placement, machine loading/unloading and assembly. These applications require fast, repeatable and articulate point-to-point movements such as palletizing, depalletizing, machine loading/unloading and assembly. ABB's SCARA family is ideal for customers requiring rapid cycle time, high precision and high reliability for their small parts assembly applications and for laboratory automation and prescription drug dispensing.

IRB 14000(YuMi)

YuMi was officially introduced to the market on April 13, 2015, making collaboration between humans and robots a reality. YuMi is a collaborative, dual arm, small parts assembly robot solution that includes flexible hands, parts feeding systems, camera-based part location and state-of-the-art robot control. YuMi is a vision of robotic co-workers, as shown in Figure 4-8.

igure 4-6 IRB 360 Robot

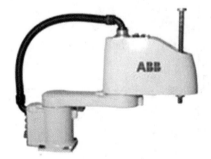
Figure 4-7 IRB 910SC Robot

Figure 4-8 YuMi Robot

Vocabulary 词汇

optimized ['ɒptɪmaɪzd]   *adj.* 最佳化的;尽量充分利用的

adjustment [ə'dʒʌstmənt]   *n.* 调整,调节;调节器

ultimately [ˈʌltɪmətli]		adv. 最后;根本;基本上
narrow [ˈnærəʊ]		adj. 精密的;度量小的 vt. 使变狭窄 vi. 变窄
tolerance [ˈtɒlərəns]		n. 公差;宽容;容忍;公差
re-engineered		(被)重组的,(被)重新调整(结构)的
flange [flændʒ]		n. 法兰;轮缘;边缘 vt. 给……装凸缘
accommodate [əˈkɒmədeɪt]		vt. 容纳;使适应;供应 vi. 适应;调解
confined [kənˈfaɪnd]		adj. 狭窄的;有限制的 v. 限制
kit [kɪt]		vt. 装备 vi. 装备
articulate [ɑːˈtɪkjʊleɪt]		vi. 用关节连接起来
depalletizing		卸垛
collaboration [kəˌlæbəˈreɪʃn]		n. 合作;勾结;通敌
reach [riːtʃ]		n. 河段;流域;范围

Notes 注释

1. functional package 功能包
2. arc welding 电弧焊
3. teaching pendant 示教器
4. loading and unloading 装载和卸载
5. consumable part 易损件
6. cycle time 周期时间;循环时间
7. hard automation 刚性自动化
8. motion control 运动控制
9. flow wrapped product 流水线包装产品
10. indexing belt 分度带
11. component placement 元件贴装
12. laboratory automation 实验室自动化
13. prescription drug dispensing 处方药分配
14. dual arm 双臂
15. feeding system 进料系统
16. camera-based part location 基于相机的工件定位
17. process application 过程应用
18. material handling 材料处理
19. reduced coating consumption 降低涂料损耗
20. single arm robot 单臂机器人
21. state-of-the-art robot control 先进的机器人控制
22. robotic co-worker 人机合作;机器人合作伙伴

Part 3 The Structure of ABB Robot

Here we focus on ABB's typical robot product IRB 120 robot, part of ABB's latest six-axis industrial robots, as shown in Figure 4-9. ABB IRB 120 robot is generally composed of three parts: robotic arm, IRC5 controller and FlexPendant.

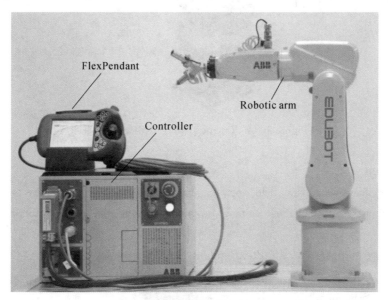

Figure 4-9 The Composition Structure of ABB IRB 120 Robot

Robotic Arm

Robotic arm is also called manipulator. It is the main mechanical arm of industrial robots; the actuator used to complete the required tasks. It is mainly composed of mechanical arm, driving device, transmission device and internal sensors. For a six-axis robot, the robotic arm mainly includes a base, a waist, an arm(boom and forearm), and a wrist, as shown in Figure 4-10.

Controller

The IRC5 controller contains all functions needed to move and control the robot. A controller consists of two modules, the control module and the drive module. The two modules are often combined in one controller cabinet.

The control module contains all the control electronics such as main computer, I/O boards, and flash memory. The control module runs all software necessary for operating the robot(that is the RobotWare system).

The drive module contains all the power electronics supplying the robot motors. An IRC5 drive module may contain nine drive units and handle six internal axes plus two or additional axes depending on the robot model.

When running more than one robot with one controller(MultiMove option), an extra drive module must be added for each additional robot. However, a single control module is

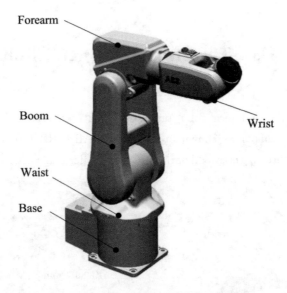

Figure 4-10 Robotic Arm of IRB 120 Robot

used.

FlexPendant

The FlexPendant is a hand held operator unit used to perform many tasks involved when operating a robot system, such as running programs, jogging the manipulator, modifying robot programs.

The surface structure of the FlexPendant is shown in Figure 4-11.

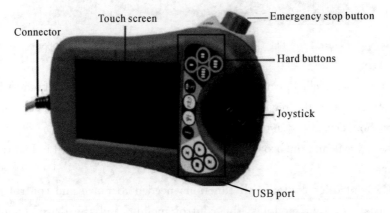

Figure 4-11 The Surface Structure of the FlexPendant

The FlexPendant consists of both hardware and software and is a complete computer itself. It is a part of IRC5, connected to the controller by an integrated cable and connector. The hot plug button option, however, makes it possible to disconnect the FlexPendant in automatic mode and continue running without it.

Vocabulary 词汇

FlexPendant 示教器

actuator [ˈæktʃueɪtə]　　　　　　　　　n. [自]执行机构；激励者

forearm [ˈfɔːrɑːm]　　　　　　　　n. 前臂 vt. 预先武装
boom [buːm]　　　　　　　　　　vt. 使兴旺 n. 繁荣;吊杆
waist [weɪst]　　　　　　　　　　n. 腰,腰部
base [beɪs]　　　　　　　　　　　n. 底部 vt. 以……作基础
wrist [rɪst]　　　　　　　　　　　n. 手腕;腕关节 vt. 用腕力移动
module [ˈmɒdjuːl]　　　　　　　　n. 模块;加载模块列表
integrated [ˈɪntɪɡreɪtɪd]　　　　　adj. 综合的;完整的;互相协调的
disconnect [ˌdɪskəˈnekt]　　　　　vt. 拆开,使分离 vi. 断开

Notes 注释
1. driving device　驱动装置
2. transmission device　传动装置
3. internal sensor　内部传感器
4. controller cabinet　控制器机柜;控制柜
5. control electronics　电子控制装置
6. main computer　主机
7. I/O board　I/O 电路板
8. flash memory　闪存
9. power electronics　电源电子设备
10. hand held operator unit　手持式操作装置
11. running program　运行程序
12. jogging the manipulator　控制机器人本体
13. emergency stop button　急停按钮
14. automatic mode　自动模式
15. drive unit　驱动单元
16. additional axes　附加轴

Part 4　Typical Robot—IRB 120

Here we focus on ABB's typical robot product IRB 120 robot, part of ABB's latest six-axis industrial robots.

IRB 120

It is ABB's smallest ever multipurpose industrial robot, it weighs just 25 kg and can handle a payload of 3 kg(4 kg for vertical wrist) with a reach of 580 mm. It is a cost-effective and reliable choice for generating high production outputs in return for low investment. A white finish Clean Room ISO 5(Class 100) version, certified by IPA, is also available.

IRB 120 is designed for robot-based flexible automation manufacturing industry(eg 3C industry), commonly used in assembly, material handling, etc, as shown in Figure 4-12.

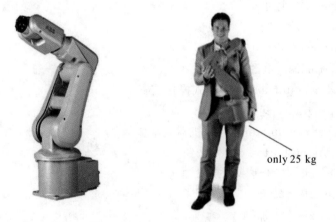

Figure 4-12　IRB 120 Robot

The specifications and features of IRB 120 robot are shown in Table 4-1 below.

Table 4-1　Specifications and Features of IRB 120

Specifications			
Variants	Reach	Payload	※Arm load
IRB 120	580 mm	3 kg	0.3 kg
Features			
Integrated Signal Supply	10 signals on wrist		
Integrated Air Supply	4 air on wrist(5 bar)		
Position Repeatability	±0.01 mm		
Robot Mounting	Any angle		
Degree of Protection	IP30		
Controllers	IRC5 compact		

※Arm load refers to the maximum total mass of the device mounted on the arm. In table 4-1, the total weight of IRB 120 robot arm can not exceed 0.3 kg.

The movement and performance of IRB 120 robot are shown in Table 4-2 below.

Table 4-2 The Movement and Performance of IRB 120

Movement		
Axis Movements	Working Range	Maximum Speed
Axis 1 Rotation	$-165°$—$+165°$	$250°/s$
Axis 2 Arm	$-110°$—$+110°$	$250°/s$
Axis 3 Arm	$-90°$—$+70°$	$250°/s$
Axis 4 Wrist	$-160°$—$+160°$	$320°/s$
Axis 5 Bend	$-120°$—$+120°$	$320°/s$
Axis 6 Turn	$-400°$—$+400°$	$420°/s$
Performance		
1 kg Picking Cycle		
※ $25 \times 300 \times 25$ mm		0.58 s
TCP Max speed		6.2 m/s
TCP Max Acceleration		28 m/s^2
Acceleration Time 0—1 m/s		0.07 s

※(1) $s_1 = 25$ mm, $s_2 = 300$ mm, $s_3 = 25$ mm.

(2) Mounted with 1 kg material on the end effector, the robot runs a circle along the path "$A \rightarrow B \rightarrow C \rightarrow B \rightarrow A$", and the time is 0.58 s, as shown on the Figure 4-13.

Figure 4-13 Running path of robot

Industrial Robot Skills Assessment Training Platform(HRG-HD1XKA)

HRG-HD1XKA industrial robot skills assessment training platform (professional edition), as shown in Figure 4-14, is a universal six-axis robot training platform, can be installed with any brand of compact six-axis robot, combined with a variety of automation mechanism, with standardized teaching modules for industrial applications, can be used for the teaching of industrial robot virtual simulation, PLC, programming, etc. It is mainly used for teaching and skills assessment of industrial robot technician.

Figure 4-14 Industrial Robot Skills Assessment Training Platform(HRG-HD1XKA)

Basic Module

The teaching plate of the basic module contains: circular groove, square groove, regular hexagonal groove, triangular groove, spline groove, and an *XOY* coordinate system, as shown in Figure 4-15. Manipulating robot and its fixture moves along the above shapes, which can be used to train a simple track teaching. The module can be used for many learnings, such as the calibration of the tool coordinate system and the workpiece coordinate system, straight teaching, arc teaching, curve teaching.

Figure 4-15　Basic Module

Vocabulary 词汇

multipurpose [ˌmʌltɪˈpəːpəs]　　　　adj. 多目标的；多种用途的
specification [ˌspesɪfɪˈkeɪʃən]　　　n. 规格；说明书；详述
variant [ˈveərɪənt]　　　　　　　　adj. 不同的；多样的 n. 变体；转化
reach [riːtʃ]　　　　　　　　　　　n. 范围；延伸；河段
mounting [ˈmaʊntɪŋ]　　　　　　　n. 装备，装配；上马
groove [gruːv]　　　　　　　　　　n. [建] 凹槽，槽；最佳状态
calibration [ˌkælɪˈbreɪʃən]　　　　n. 校准；刻度；标度

Notes 注释

1. flexible automation　柔性自动化
2. integrated signal supply　集成信号接口
3. integrated air supply　集成气路接口
4. position repeatability　重复定位精度
5. robot-based　基于机器人的

6. degree of protection   防护等级
7. the movement and performance   运动范围及性能
8. axis movement   轴运动
9. working range   工作范围
10. maximum speed   最大速度
11. picking cycle   拾料节拍
12. acceleration time   加速时间
13. training platform   实训台
14. teaching module   教学模块
15. virtual simulation   虚拟仿真
16. track teaching   轨迹示教
17. tool coordinate system   工具坐标系
18. workpiece coordinate system   工件坐标系
19. straight teaching   直线示教
20. arc teaching   圆弧示教
21. curve teaching   曲线示教

Unit 5　KUKA Robot

Part 1　KUKA and KUKA Robot

KUKA

KUKA is a German manufacturer of industrial robots and solutions for factory automation. The corporate history of KUKA began in 1898 with Johann Joseph Keller and Jakob Knappich in Augsburg and KUKA is created from the first letters of "Keller und Knappich Augsburg". For more than 100 years, KUKA has stood for ideas and innovations that have made it successful worldwide. More than 12 000 KUKA colleagues worldwide are developing the intelligent, robot-based automation solutions of tomorrow—creative, integrating and effective. It is called "Orange Intelligenz". At the interface between the virtual and real worlds, "Orange Intelligenz" is the creative, integrating and effective power of KUKA.

KUKA

Today, KUKA is one of the world's leading suppliers of automation solutions. KUKA stands for innovations in automation and is a driver of Industry 4.0.

KUKA Robot

As a specialist in the field of robotics and automation technology, KUKA Robotics is one of the leading manufacturers of industrial robots. With its wide selection of robots, KUKA covers virtually all payload ranges and robot types, and sets standards in the field of human-robot collaboration(HRC).

1. Pioneer in the Industrial Robots Sector

KUKA has been manufacturing industrial robots for more than 40 years. The cornerstone for today's Industry 4.0 was laid as far back as in the 1970s with its first industrial robot FAMULUS. In 1996, KUKA Robotics took another quantum leap in the development of industrial robots: that year it launched the first PC-based controller, developed by KUKA. This was the dawn of the "real" mechatronics era, characterized by the precise interaction of software, controller and mechanical systems. Nowadays, KUKA

Robotics offers a wide product range for various industries and tailor-made automation solutions. KUKA industrial robots improve the quality of the products and reduce the requirements for costly materials and limited energy resources.

At KUKA Robotics, its vision is to establish the industrial robot as an intelligent assistant to humans during manufacturing: humans and robots work hand in hand, ideally complementing each other with their respective skills.

2. Products and Industries

KUKA Robotics supplies industrial robots which are perfectly tailored to the applications of customers. From the actual robot itself and the controller all the way to the appropriate software, customers from a diverse range of industries benefit from innovative technologies and sophisticated engineering.

KUKA Robotics can offer the following product spectrum: six-axis robots with virtually any reach and payload capacity; robots for the resistant to heat, dirt and water; robots for the food and pharmaceutical industries; clean room versions; palletizing robots; welding robots; shelf-mounted robots and high-accuracy robots. And the solutions of KUKA Robotics are implemented in the following industries and customers in particular: arc welding and many other welding processes, machine tools industry, foundry and forging industry, plastics industry, electronics industry, food industry, automotive suppliers and automotive manufacturers(as shown in Figure 5-1).

Figure 5-1   Automation in the Automotive Industry

Vocabulary 词汇

creative [kriˈeɪtɪv]		adj. 创造性的,创新的
integrate [ˈɪntɪɡreɪt]		v. 集成化;综合化
effective [ɪˈfektɪv]		adj. 有效的;起作用的;实际的
forging [ˈfɔːdʒɪŋ]		n. 锻造;锻件 v. 锻造;打制
innovation [ˌɪnəˈveɪʃən]		n. 创新,革新;新方法
specialist [ˈspeʃəlɪst]		n. 专家;专门医师 adj. 专家的;专业的
virtually [ˈvɜːtʃuəli]		adv. 事实上,几乎;实质上

tailor-made ['teɪlə'meɪd]   adj. 特制的;裁缝制的
complement ['kɒmplɪment]   n. 补语;余角 vt. 补足,补助

Notes 注释
1. factory automation   工厂自动化
2. orange intelligenz   橙色智能
3. industry 4.0   工业 4.0
4. human-robot collaboration   人机协作
5. quantum leap   质的飞越;重大突破
6. PC-based controller   基于 PC 的控制器
7. energy resource   能源资源
8. payload capacity   负载能力;有效载荷
9. precise interaction   精准操作
10. intelligent assistant   智能助手
11. shelf-mounted robot   架装式机器人

Part 2　KUKA Robot Product Series

KUKA offers a comprehensive range of industrial robots. Customers will always find the right robots, no matter how challenging the application is. The following is a brief description of the main models of KUKA robots (the specific parameters of the specifications are from KUKA officially announced data).

KR 16

KR 16 is one of the most versatile six-axis industrial robots of KUKA, as shown in Figure 5-2. It can be expanded easily and is available in various combinations. Its jointed-arm kinematic system makes it the perfect partner for all point-to-point and continuous-path controlled tasks in the low payload category.

KR 700 PA

KR 700 PA is the ideal palletizer for heavier payloads, as shown in Figure 5-3. With its enormous reach and compact design, it adapts perfectly to every application. No matter what you use it for, optimal cycle time is assured.

Figure 5-2　KR 16 Robot　　　　Figure 5-3　KR 700 PA Robot

KR AGILUS sixx

KR AGILUS sixx is a KUKA compact six-axis robot that is designed for particularly high working speed, as shown in Figure 5-4. Different versions, installation positions, reaches and payloads transform the small robot into a precision artist. Irrespective of the installation position—whether on the floor, ceiling or wall—it achieves utmost precision in confined spaces thanks to its integrated energy supply system and service-proven KR C4 compact controller.

KR CYBERTECH ARC nano

KR CYBERTECH ARC nano product family is optimized for continuous-path applications such as arc welding and the application of adhesives and sealants, as shown in Figure 5-5. The industrial robots offer ideal performance combined with a high power density—for maximum economy at low cost. The 50-millimeter hollow-shaft wrist is a future-oriented innovation, the hollow axis allows reduced main axis motion with short cycle time and utmost precision of movement.

igure 5-4　KR AGILUS sixx Robot　　　　igure 5-5　KR CYBERTECH ARC Robot

KR QUANTEC pro

KR QUANTEC pro has been optimized to work with payloads between 90 and 120 kg, as shown in Figure 5-6. It is compact, powerful and extremely precise—that opens up the possibility of new, innovative cell concepts for Industry 4.0. Its slender wrist and the reduced interference contour allow for extremely high precision and speed. As a result, it is perfectly suited to spot welding, soldering and handling tasks.

LBR iiwa

LBR iiwa is the world's first series-produced sensitive robot, and it's a human-robot-collaboration robot, as shown in Figure 5-7. This signals the beginning of a new era in industrial, sensitive robotics and lays the foundation for innovative and sustainable production processes. For the first time, humans and robots can work together on highly sensitive tasks in close cooperation. This opens up the possibility of new applications and the way is paved for greater cost-effectiveness and utmost efficiency. LBR iiwa robot is available in two versions with payload capacities of 7 and 14 kg.

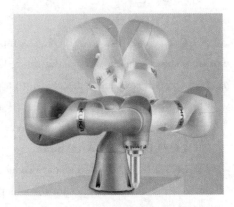

Figure 5-6　KR QUANTEC pro Robot　　　　Figure 5-7　LBR iiwa Robot

Vocabulary 词汇

powerful ['paʊəfəl]	adj. 强大的 adv. 很；非常
soldering ['sɒldərɪŋ]	n. 焊接 adj. 用于焊接的 v. 焊接
challenging ['tʃælɪndʒɪŋ]	v. 要求；质疑；反对；向……挑战
combination [ˌkɒmbɪ'neɪʃən]	n. 结合；组合；联合
enormous [ɪ'nɔːməs]	adj. 庞大的，巨大的
adhesive [əd'hiːsɪv]	n. 黏合剂；胶黏剂 adj. 黏着的；带黏性的
sealant ['siːlənt]	n. 密封剂
slender ['slendə]	adj. 细长的；苗条的

Notes 注释

1. six-axis industrial robot 六轴工业机器人
2. jointed-arm kinematic system 关节运动系统
3. continuous-path 连续路径
4. hollow axis 空心轴
5. power density 功率密度
6. installation position 安装位置
7. energy supply system 能源供应系统
8. interference contour 干扰轮廓
9. spot welding 点焊
10. brief description 项目简介
11. continuous-path controlled 轨迹控制
12. low payload 低负载区
13. series-produced adj. 量产的

Part 3　The Structure of KUKA Robot

Here we focus on KR QUANTEC pro robot in the high payload range, part of KUKA's six-axis industrial robots. KUKA KR QUANTEC pro robot is generally composed of three parts: manipulator, KR C4 controller and smartPAD, as shown in Figure 5-8.

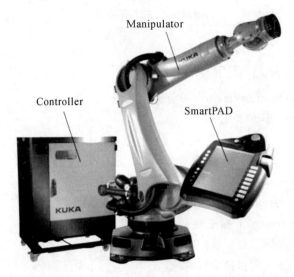

Figure 5-8　The Composition Structure of KR QUANTEC pro Robot

Manipulator

Manipulator is also called robot body. It is the main mechanical arm of industrial robots; the actuator used to complete the required tasks. It is mainly composed of robotic arm, driving device, transmission device and internal sensors. For a six-axis robot, the robotic arm mainly includes a base, a waist, two arms(link arm and forearm), and a wrist, as shown in Figure 5-9.

Controller

KR C4 controller is a pioneer for the automation of today and tomorrow. It reduces costs in integration, maintenance and servicing. At the same time, the long-term efficiency and flexibility of the systems are increased thanks to common, open industry standards. The KR C4 software architecture integrates Robot Control, PLC Control, Motion Control(for example, KUKA. CNC) and Safety Control. All controllers share a database and infrastructure. With five variants(KR C4 compact, KR C4 smallsize -2, KR C4, KR C4 midsize and KR C4 extended), KR C4 can be optimally integrated into the automation environment. Requirements for stackability, protection against dust, humidity and other influences can thus be taken into account.

KR C4 consists of the following components, as shown in Figure 5-10.

SmartPAD(Teach Pendant)

The smartPAD is a hand held operator unit used to perform almost all of the tasks

Figure 5-9   Robotic Arm of KR QUANTEC pro Robot

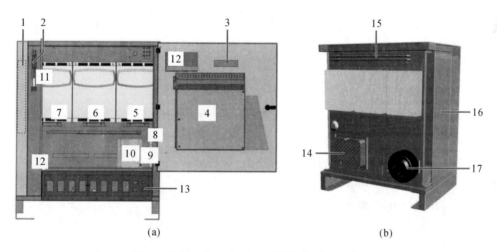

Figure 5-10   Constitution of KR C4 Controller

(a) Front View   (b) Rear View

1—Mains filter; 2—Main switch; 3—Controller System Panel(CSP); 4—Control PC;
5—Drive power supply(drive controller for axes 7 and 8, optional); 6—Drive controller for axes 4 to 6;
7—Drive controller for axes 1 to 3; 8—Brake filter; 9—Cabinet Control Unit(CCU);
10—Safety Interface Board(SIB)/Extended SIB; 11—Transient limiter; 12—Batteries(positioning depending on variant);
13—Connection panel; 14—Low-voltage power supply unit; 15—Brake resistor; 16—Heat exchangers; 17—External fan

involved when operating a robot of KUKA.

The KUKA smartPAD is designed to master even complex operating tasks easily. It can be deployed universally and is easy to operate, even for inexperienced users. It's able to operate all KUKA robots with a KR C4 controller.

The surface structure of the smartPAD is shown in Figure 5-11.

The smartPAD consists of both hardware and software and is a complete computer itself. It is connected to the controller by an integrated cable and connector.

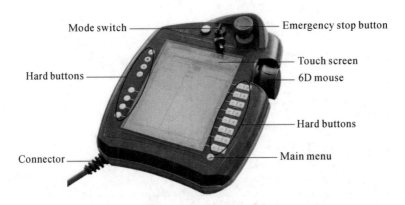

Figure 5-11　The Surface Structure of the SmartPAD

Vocabulary 词汇

integration [ˌɪntɪˈgreɪʃən]　　　　n. 集成；综合
infrastructure [ˈɪnfrəˌstrʌktʃə]　　n. 基础设施；公共建设；下部构造
humidity [hjuːˈmɪdəti]　　　　　n. [气象]湿度；湿气
hardware [ˈhɑːdweə]　　　　　　n. 计算机硬件；五金器具

Notes 注释

1. smartPAD　示教器
2. transmission device　传动装置
3. internal sensor　内部传感器
4. open industry standard　开放式工业标准
5. protection against dust　防尘
6. protection against humidity　防潮
7. mains filter　电源滤波器
8. Controller System Panel(CSP)　控制器系统面板(CSP)
9. drive power supply　驱动电源
10. drive controller　伺服驱动器
11. brake filter　制动过滤器
12. safety interface board　安全接口面板
13. low-voltage power supply unit　低压供电单元
14. heat exchanger　热交换器
15. brake resistor　制动电阻
16. integrated cable　集成电缆；综合布线
17. main mechanical　机械主体
18. pioneer for the automation　自动化先锋
19. software architecture　软件架构
20. drive controller for axes　伺服驱动器轴
21. transient limiter　瞬态抑制器

Part 4　Typical Robot—KR 6 R700

Here we focus on one typical robot product—KR 6 R700 sixx robot, part of KUKA's small six-axis industrial robots.

KR 6 R700 sixx

KR 6 R700 sixx is a KUKA's compact six-axis robot that is designed for particularly high working speed, as shown in Figure 5-12. Irrespective of the installation position—whether on the floor, ceiling or wall—it achieves utmost precision in confined spaces thanks to its integrated energy supply system and service-proven KR C4 compact controller.

Figure 5-12　KR 6 R700 sixx Robot

Its features are as follows.

(1) Minimum cycle time. The KR 6 R700 sixx has six axes and is consistently rated for particularly high working speed. At the same time, it offers high precision.

(2) Space-saving integration. Low space requirements and the choice between installation on the floor, ceiling or wall make the KR 6 R700 sixx extremely adaptable.

(3) Integrated energy supply system. It is routed internally in the robot, thereby saving space. It includes EtherCAT/EtherNet(bus cable), three 5/2-way valves(compressed air), direct air line and inputs/outputs.

(4) KUKA.SafeOperation. The robot sets standards in safety. Only can they offer the KUKA.SafeOperation functionality, which radically simplifies the effective cooperation of humans and machines.

The specifications and features of the KR 6 R700 sixx robot are shown in Table 5-1 below.

Table 5-1 The Specifications and Features of KR 6 R700 sixx

Specifications		
Variants	Max. reach	Max. payload
KR 6 R700 sixx	706.7 mm	6 kg
Features		
Pose repeatability	±0.03 mm	
Mounting position	Floor, ceiling, wall	
Robot footprint	209 mm × 207 mm	
Weight(excluding controller), approx.	50 kg	
Ambient temperature	+5 to +45 ℃	
Protection rating	IP 54	
Controller	KR C4 compact	

The movement of KR 6 R700 sixx robot is shown in Table 5-2 below.

Table 5-2 The Movement of KR 6 R700 sixx

Movement		
Axis	Range of motion	Speed with rated payload
Axis 1(A1)	−170°—+170°	250°/s
Axis 2(A2)	−190°—+45°	250°/s
Axis 3(A3)	−120°—+156°	250°/s
Axis 4(A4)	−185°—+185°	320°/s
Axis 5(A5)	−120°—+120°	320°/s
Axis 6(A6)	−350°—+350°	420°/s

Industrial Robot Skills Assessment Training Platform(HRG-HD1XKA)

HRG-HD1XKA industrial robot skills assessment training platform (professional edition), as shown in Figure 5-13, is a universal six-axis robot training platform. It can be installed with any brand of compact six-axis robot, combined with a variety of automation mechanism, with standardized teaching modules for industrial applications, and can be used for the teaching of industrial robot virtual simulation, PLC, programming, etc. It is mainly used for teaching and skills assessment of industrial robot technician.

Laser Engraving Module

The laser of the module runs along the track of the engraving panel and it can be used to learn laser engraving applications, to achieve the skillful application of the basic functions and I/O signal configuration, as shown in Figure 5-14.

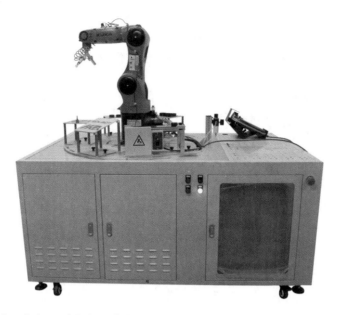

Figure 5-13 Industrial Robot Skills Assessment Training Platform(HRG-HD1XKA)

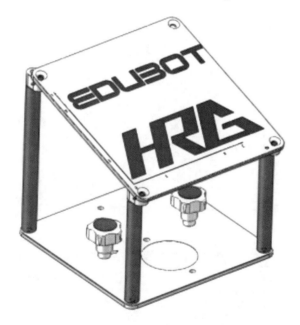

Figure 5-14 Laser Engraving Module

Vocabulary 词汇

laser ['leɪzə]   n. 激光
engraving [ɪn'greɪvɪŋ]   n. 雕刻；雕刻术 v. 在……上雕刻；给……深刻的印象
configuration [kənˌfɪɡjʊ'reɪʃən]   n. 配置；结构；外形

Notes 注释

1. compact six-axis robot   紧凑式六臂机器人

2. utmost precision　最高精度

3. space-saving integration　节省空间的集成

4. integrated energy supply system　集成式能源供应系统

5. etherNet　以太网

6. bus cable　总线电缆

7. 5/2-way valve　二位五通阀

8. compressed air　压缩空气

9. cooperation of humans and machines　人机协作

10. engraving panel　雕刻面板

11. minimum cycle time　最短循环时间

12. direct air line　直接气路

Unit 6　YASKAWA Robot

Part 1　YASKAWA and YASKAWA Robot

YASKAWA

The YASKAWA Electric Corporation is the world's largest manufacturer of AC inverter drives, servo and motion control, and robotic automation systems. The company was founded in 1915, and its head office is located in Japan. In the late 1960s, YASKAWA Electric Corporation led the world in putting forward the concept of "mechatronics". This concept evolved when YASKAWA combined its customers' machineries with YASKAWA's electric products to create superior quality and function.

YASKAWA

Since 1915, YASKAWA has served the world's needs for products to improve global productivity through automation. YASKAWA started the mid-term business plan "Dash 25" in 2016 as the first step towards realizing the long-term business plan "Vision 2025".

YASKAWA Robot

Since the debut of the all-electric industrial robot "MOTOMAN" in Japan in 1977, YASKAWA has taken a leading role in the world's industrial robot market. Starting with arc welding for automobile production, for which YASKAWA remains a market leader, YASKAWA robots have assumed an active role in every industrial field including welding, packaging, assembly, coating, cutting, material handling and general automation, as well as handling and transport in clean rooms for liquid crystal, organic EL displays and semiconductor manufacturing, as shown in Figure 6-1.

As one solution to the labor shortage in recent years, the scope of YASKAWA robot applications is extending to the food industry and other markets where there has been little practice in robotic automation. YASKAWA promotes the creation of new markets for robots and the acceleration of open innovation.

Recently, YASKAWA robots have expanded in appeal and presence in not only industrial but also other fields by advancing speed, accuracy improvement, response to complex movements, and strengthening of safety functions toward coexistence with

Figure 6-1　Auto Spot Welding

humans. The development of YASKAWA robots actively participating in fields closer to humans will continue to advance in the future.

Vocabulary 词汇

mechatronics [ˌmekəˈtrɒniks]	n. 机电一体化；机械电子学
debut [ˈdeɪbjuː]	n. 初次登台；首次出场 vi. 初次登台
welding [ˈweldɪŋ]	adj. 焊接的 n. 焊接 v. 焊接
packaging [ˈpækɪdʒɪŋ]	n. 包装；包装业
assembly [əˈsembli]	n. 装配；集会；集合
coating [ˈkəʊtɪŋ]	n. 涂层
cutting [ˈkʌtɪŋ]	n.[机]切断 adj. 锋利的 v. 削减
display [dɪˈspleɪ]	n.[电子]显示器 v.[电子]显示

Notes 注释

1. AC inverter drive　交流变频器驱动
2. robotic automation system　机器人自动化系统
3. customers' machinery　客户机械
4. electric product　电气产品
5. business plan　商业计划
6. long-term business plan　长期商业计划
7. all-electric industrial robot　全电动工业机器人
8. general automation　自动化生产
9. organic EL display　有机EL显示器
10. semiconductor manufacturing　半导体制造
11. labor shortage　劳动力短缺

Part 2 YASKAWA Robot Product Series

YASKAWA has successively commercialized and marketed optimum robots for various uses since 1977, centering on arc welding, one of its areas of expertise, and including spot welding, handling, assembly, painting, transfer of liquid crystal panels, transfer of semiconductor wafers and so on. The following is a brief description of the main models of YASKAWA robots (the specific parameters of the specifications are from YASKAWA officially announced data).

MH5S II

MH5S II is a compact, high-speed six-axis robot that offers superior performance in a variety of applications such as packaging, material handling, machine tending and dispensing, where versatility is required, as shown in Figure 6-2.

MH5S II features a 706 mm reach and offers the widest work envelope in its class. All shafts are low power output and don't need to set the security fence. It can effectively reduce the collision of robot and peripheral devices. MH5S II is controlled via the DX200 robot controller, which is capable of controlling up to 8 robots from the same controller. Thanks to the compact design of this robot and built-in collision detection features, multiple robots can be easily accommodated in a single production facility.

MPP3S

MPP3S is a four-axis high-speed robot with parallel kinematic system and it is combined with the speed of the delta design with a high payload capacity and a large working range, as shown in Figure 6-3, and is suitable for high speed and high precision of palletizing, pick-up and packaging.

Figure 6-2 MH5S II Robot

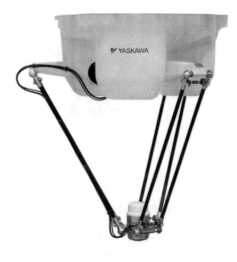
Figure 6-3 MPP3S Robot

It can be installed in a small space. As the best in its class, it also enables high-speed handling with a large motion range and improves productivity. Cycle time can be reduced with shorter suction time in addition to high-speed robot movements. The hollow structure of the robot (first in the industry) enables the installation of air valves inside the parallel link arms. The structure shortens the length of piping drastically, which achieves a reduction in the cycle time. MPP3S is also designed to maintain high level of cleanliness for food handling.

MPL300 II

The versatile and powerful four-axis robot MPL300 II provides high performance in box palletizing, case palletizing and many further logistical tasks for end-of-line or distribution center automation, as shown in Figure 6-4. Its extensive vertical reach of 3 024 mm combined with 3 159 mm horizontal reach enables high palletizing loads (Payload: 300 kg). Internally routed airlines and cables from base to end-of-arm tool maximize system reliability.

EPX2800

The flexible, high-performance MOTOMAN painting robot EPX2800 increases finishing quality, consistency and throughput, while dramatically lowering operating costs and decreasing wasted materials. EPX2800 robot is ideal for automotive and other industrial coating applications, as shown in Figure 6-5. It offers superior performance and creates smooth, consistent finish with outstanding efficiency for painting and dispensing applications. This robot features a hollow wrist design and is ideal for painting contoured parts such as interior/exterior surfaces. It is well suited for mounting spray equipment applicators. Interference between the hose and parts/fixtures is avoided thus ensuring optimum cycle time and robot reach/access.

Figure 6-4 MPL300 II Robot   Figure 6-5 EPX2800 Robot

MYS450F

MYS450F is a four-axis SCARA robot that offers high speed in a compact form that requires minimal installation space, as shown in Figure 6-6. MYS450F offers superior performance in applications such as assembly, small part handling, case packing and lab automation. The SCARA robot easily integrates with existing robot applications to expand current automated processes. It is ideal for large, multi-process systems requiring pick-and-place capability.

SIA10D

SIA10D is a lean and powerful seven-axis single-arm robot ideal for automating operations such as assembly, inspection, machine tending and handling, as shown in Figure 6-7. It's revolutionary design with high wrist performance and fully integrated supply cables, enables the SIA10D to work in confined spaces with an amazing freedom of movements. With a minimal footprint and high motion flexibility, the SIA10D can be positioned out of the normal working area (i.e. floor-, ceiling-, wall-, incline-or machine-mounted) without limiting the motion range of any axis. To save valuable floor space the SIA10D can be mounted between machines, which also provides open access to the machines for maintenance, adjustment or testing.

Figure 6-6　MYS450F Robot　　　　igure 6-7　SIA10D Robot

Vocabulary 词汇

compact [kəmˌpækt]　　　　　*adj.* 紧凑的,紧密的 *vt.* 使简洁
flexible [ˈfleksɪbəl]　　　　　*adj.* 灵活的;柔韧的;易弯曲的

Notes 注释

1. low power output　低功率输出
2. security fence　安全性防护
3. built-in collision detection feature　内置碰撞检测功能
4. speed of the delta design　速度增量设计
5. shorter suction time　较短的吸入时间
6. hollow structure　中空结构
7. piping drastically　从动臂管
8. high level of cleanliness　高度清洁
9. high palletizing load　高码垛负载
10. lab automation　实验室自动化
11. multi-process system　多进程系统

Part 3　The Structure of YASKAWA Robot

Here we focus on one high performance robot product—MH12 robot, part of YASWAKA's six-axis industrial robots, as shown in Figure 6-8. YASKAWA MH12 robot is generally composed of three parts: manipulator, DX200 controller and programming pendant.

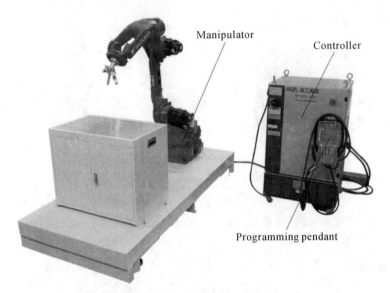

Figure 6-8　The Composition Structure of YASKAWA MH12 Robot

Manipulator

Manipulator is also called robot body. It is the main mechanical arm of industrial robots; the actuator used to complete the required tasks. It is mainly composed of robotic arm, driving device, transmission device and internal sensors. For a six-axis robot, the robotic arm mainly includes a base, a waist, two arms (upper arm and forearm) and a wrist, as shown in Figure 6-9.

Controller

YASKAWA's new DX200 controller features robust PC architecture and system-level control for robotic work cells.

It uses patented multiple robot control technologies, for example, I/O devices and communication protocols, furthermore it provides built-in ladder logic processing including 4 096 I/O addresses, a variety of fieldbus network connections, a high-speed E-server connection and I/F panel which shows the human machine interface (HMI) on the pendant. It often eliminates the need for separate PLC and HMI and delivers significant cost savings at system level, while decreasing work cell complexity and improving overall reliability. Dynamic interference zones protect the robot arm and provide advanced collision avoidance. The Advanced Robot Motion (ARM) control provides high performance, best-in-class path

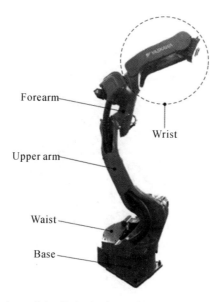

igure 6-9 Robotic Arm of MH12 Robot

planning and dramatically reduces teaching time. It supports coordinated motion with multiple robots or other devices. A small, light weight Windows® CE programming pendant features colour touch screen with multiple window display capability. Programming features are designed to use minimum number of keystrokes and are facilitated by new function packages and more than 120 functions. Furthermore it conserves the power consumption from 38%~70% depending on application and robot size. It is available with the optional category 3 Functional Safety Unit(FSU) and allows an establishment of 32 safety units and up to 16 tools.

Its key features are as follows:

(1) specific application function package including more than 120 functions;
(2) optional category 3 Functional Safety Unit(FSU);
(3) high productivity;
(4) low integration cost;
(5) integrated cell control capabilities;
(6) high reliability and energy efficiency;
(7) easy maintenance;
(8) simple programming;
(9) convenient compact flash slot and USB port facilitate memory backups.

Programming Pendant(Teach Pendant)

The programming pendant is a hand held operator unit used to perform almost all of the tasks involved when operating a robot of YASKAWA. The main role of programming pendant is to teach the manipulator movement.

The surface structure of the programming pendant is shown in Figure 6-10.

The programming pendant consists of both hardware and software and is a complete computer itself. It is connected to the controller by an integrated cable and connector.

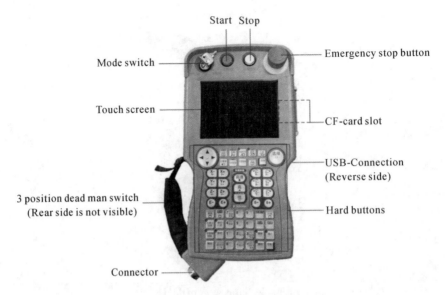

Figure 6-10　The Surface Structure of the Programming Pendant

Vocabulary 词汇

manipulator [məˈnɪpjʊleɪtə]　　　n. 操纵器,调制器;操作者
maintenance [ˈmeɪntənəns]　　　n. 维护;维修;维持,保持;保养,保管
keystroke [ˈkiːstrəʊk]　　　v. 击键 n. 打键

Notes 注释

1. DX200 controller　DX200 控制器
2. programming pendant　示教器
3. main mechanical　主要机械
4. patented multiple robot control technology　机器人控制技术专利
5. communication protocol　通信协议
6. built-in ladder logic processing　内置梯形图逻辑处理
7. fieldbus network connection　现场总线网络连接
8. E-server connection　电子服务器连接
9. human machine interface　人机界面
10. advanced collision avoidance　高级碰撞避免
11. high performance　高性能
12. best-in-class path planning　一流的路线规划
13. teaching time　教学时间
14. colour touch screen　彩色触摸屏
15. function package　函数包
16. safety unit　安全单元
17. integrated cell control capability　集成单元控制功能

18. flash slot 闪存槽
19. computer numerical control 计算机数控系统
20. numerical control 数字控制
21. servo control 伺服控制
22. integrated vision system 综合视觉系统
23. intelligent robot 智能机器人

Part 4　Typical Robot—MH12

Here we focus on a typical robot product—MH12 robot, part of YASKAWA's high speed six-axis industrial robots.

MH 12

The flexible, high speed six-axis robot MOTOMAN MH12 with its payload up to 12 kg is newly added to the YASKAWA product line MOTOMAN MH series.

By realising the highest motion performance in its class, YASKAWA is contributing to the productivity development. The hollow shaft structure is adopted to the upper arm. By storing the cables in the arm, operation restriction due to the cable interference is reduced. Therefore, maintainability such as simplifying the robot teaching operation or eliminating the cable disconnection problems is improved. Stream-lined structure adopted to the new type arc welding purpose robot is also applied to this multipurpose-applicable robot. And this adoption contributes to reducing the interference area between the jigs and work pieces. Superior performance can be delivered in handling operations which require, for example the rotation of large-sized work pieces. The servo-float function enables, for example in plastic injection moulding machines, to reduce the handling of work pieces which are pushed back by the ram, including the arm of the robot.

The specifications and features of MH12 robot are shown in Table 6-1 below.

Table 6-1　The Specifications and Features of MH12

Specifications		
Variants	Max. Working Range	Max. Payload
MH12	1 440 mm	12 kg
Features		
Repeat. pos. accuracy	±0.08 mm	
Temperature	0—+45℃	
Humidity	20%～80%	
Weight	130 kg	
Power supply, average	1.5 kVA	

The movement of the MH12 robot is shown in Table 6-2 below.

Table 6-2　The Movement of MH12

Axis	Movement	
	Maximum Motion Range	Maximum Speed
S (Axis 1)	$-170°-+170°$	$220°/s$
L (Axis 2)	$-90°-+155°$	$200°/s$
U (Axis 3)	$-175°-+240°$	$220°/s$
R (Axis 4)	$-180°-+180°$	$410°/s$
B (Axis 5)	$-135°-+135°$	$410°/s$
T (Axis 6)	$-360°-+360°$	$610°/s$

Industrial Robot Skills Assessment Training Station(HRG-HD1XKD)

HRG-HD1XKD industrial robot skills assessment training station(standard edition), as shown in Figure 6-11, sets the training and assessment of industrial robot in one. It is a practical, extensible teaching equipment of industrial robot, with a variety of modules. It can be installed with any brand of industrial six-axis robot, integrated basic module such as TCP calibration, etc, equipped with laser engraving, workpiece handling, assembly and other teaching module for industrial application, the station is conducive to learn the operation skills of industrial robot from the shallower to the deeper. The training station can also be equipped with the teaching module of other knowledge for industrial robot, such as visual systems, PLC programming systems, etc. It is ideal for teaching and skills assessment of industrial robot technician.

Welding Module

The emulational welding torch forms the welding trace along the point to be welded. In order to demonstrate the welding function well, the robot should deal with the pose change of the welding torch at the corner position, and control the speed and pose during the whole welding process, as shown in Figure 6-12.

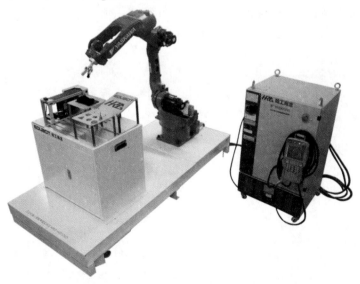

Figure 6-11　Industrial Robot Skills Assessment Training Station(HRG-HD1XKD)

Figure 6-12　Welding Module

Vocabulary 词汇

flexible ['fleksɪbəl]　　　*adj.* 灵活的；柔韧的；易弯曲的

Notes 注释

1. cable interference　电缆干扰
2. eliminating the cable disconnection　消除电缆断裂
3. arc welding purpose robot　弧焊机器人
4. work piece　工件
5. servo-float function　伺服浮动功能
6. injection moulding machine　注塑机

Unit 7　FANUC Robot

Part 1　FANUC and FANUC Robot

FANUC

FANUC has consistently pursued the automation of factories since 1956, when it succeeded in the development of the servo mechanism for the first time in the Japanese private sector.

FANUC

FANUC is a group of companies, principally FANUC Corporation of Japan, which provides automation products and service such as robotics and computer numerical control systems. FANUC is one of the largest makers of industrial robots in the world. FANUC had its beginning as part of Fujitsu developing early numerical control (NC) and servo systems. The company name is an acronym for Fuji Automatic Numerical Control.

In 1972, the Computing Control Division became independent and FANUC Ltd. was established. The company's clients include US and Japanese automobile and electronics manufacturers. FANUC has over 240 joint ventures, subsidiaries, and offices in over 46 countries. It is the largest maker of CNC controls by market share with 65% of the global market and is the leading global manufacturer of factory automation systems.

FANUC Robot

Since the first FANUC robot was developed in 1974, FANUC is committed to leading the robot technology and innovation. Now FANUC is the only company in the world to do the robot by a robot, and is the only enterprise to provide integrated vision system of the robot in the world, meanwhile it is also the only company in the world to provide intelligent robots and intelligent machines.

The series of FANUC robotic products is as many as 240 species, payload is from 0.5 kg to 2.3 tons. Its products are widely used in different production processes, such as assembly, handling, welding, casting, painting, palletizing etc, to meet the different needs of customers, as shown in Figure 7-1.

In June 2008, FANUC, which installed capacity was above 200 000, became the first robot manufacturer in the world. Moreover, the installed capacity of FANUC robots in the world has been more than 250 000 in 2011 and its market share has ranked first since then.

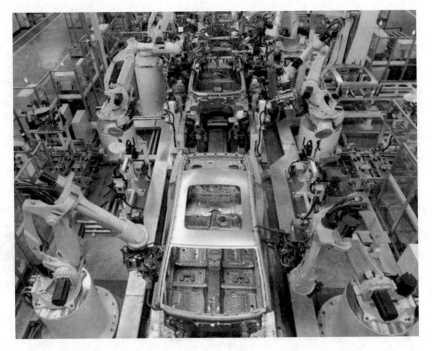

Figure 7-1   Industrial Applications of FANUC Robots

Vocabulary 词汇

servo [ˈsɜːvəʊ]   n. 伺服；伺服系统；随动系统
intelligent [ɪnˈtelɪdʒənt]   adj. 智能的；理解力强的
assemble [əˈsembəl]   vt. 聚集；装配；收集 vi. 聚集
handling [ˈhændlɪŋ]   n. 处理 adj. 操作的 v. 负责；对待
casting [ˈkɑːstɪŋ]   n. 铸造；铸件 v. 铸造；投掷
painting [ˈpeɪntɪŋ]   n. 绘画；着色 v. 绘画；涂色于
palletizing [ˈpælətaɪzɪŋ]   n. 码垛堆积；托盘包装 v. 把……装在货盘上

Notes 注释

1. computer numerical control system   计算机数控系统
2. servo system   伺服系统
3. computing control   计算控制
4. integrated vision system   集成视觉系统

Part 2 FANUC Robot Product Series

In order to meet the market demand, FANUC has developed a series of industrial robot products since 1974, such as handling robots, welding robots, assembly robots, painting robots and so on. The following is a brief description of the main models of FANUC robots(the specific parameters of the specifications are from FANUC officially announced data).

M-2iA/3S

M-2iA/3S series are the medium sized parallel link robots for high speed picking and assembly. M-2iA/3S robot is a total 3 axes robot with single axis rotation at wrist with 3 kg payload, as shown in Figure 7-2. It is suitable for high speed picking of plenty work pieces on the conveyor while aligning direction of each work piece. Moreover, it can withstand a high pressure stream cleaning with IP69K of enclosed mechanical unit and contribute to the automation in the food market. Latest intelligent functions can be applied to M-2iA/3S robot using vision sensor (iRVision) or force sensor.

M-410iB/160

M-410iB/160 series are the intelligent palletizing robots which contribute for robotization of palletizing system. M-410iB/160 robot is a high speed type with 160 kg payload, as shown in Figure 7-3. Customer can use the latest intelligent function of iRVision and ROBOGUIDE, such as detection of height of workpieces, discrimination of size or kind, visual inspection and depalletizing the load on a received pallet.

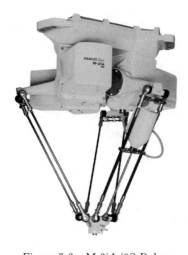

Figure 7-2 M-2iA/3S Robot

Figure 7-3 M-410iB/160 Robot

ARC Mate 120iC

ARC Mate 120iC is a torch cable integrated arc welding robot whose payload is 20 kg, as shown in Figure 7-4. High rigidity arm and advanced servo technology enable to increase its acceleration performance. This decreases cycle time and realizes high productivity. The unique drive mechanism in wrist axes realizes the slim arm in torch cable integrated robot. Thanks to stable hand cable management, off-line teaching function by ROBOGUIDE reduces teaching time of the robot.

M-900iA/260L

M-900iA/260L is a heavy payload and long arm type robot which has a 260 kg payload capacity and 3.1 m maximum reach, as shown in Figure 7-5. It enables faster long-distance transfer of large workpieces, such as transferring car body and setting large casting on machining fixture. The wrist can be used safely in adverse environments because it has IP67-equivalent resistance to environmental conditions(dust and dips).

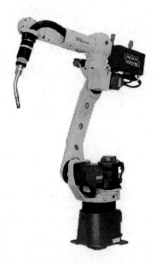

Figure 7-4　ARC Mate 120iC Robot

Figure 7-5　M-900iA/260L Robot

P-250iB

P-250iB is one of the most flexible and advanced six-axis painting robots available in the market, as shown in Figure 7-6. The P-250iB is adaptable in its mounting configuration to accommodate the most demanding of painting applications. The "Open Architecture" design allows the use of standard FANUC robotics integrated process equipment and streamlines easy integration of third party application equipment. The optional on-arm pneumatic enclosure provides the ability to mount FANUC robotics' approved components crucial to overall performance and paint quality within the inner arm.

CR-35iA

CR-35iA is a 35 kg payload collaborative robot that can work in cooperation with human operator, as shown in Figure 7-7. The robot and human operator can work together within a shared workspace, for example, heavy workpieces transfer and parts assembly, without safety fence. This robot stops safely when it touches human operator. Safe and gentle looking green soft cover reduces an impact force and prevents human operator from being pinched. CR-35iA is certified to meet the requirements of international standard ISO 10218-1. It is designed with the same high reliability as conventional robots.

Figure 7-6　P-250iB Robot

Figure 7-7　CR-35iA Robot

Vocabulary 词汇

aligning [əˈlaɪnɪŋ]　　　　　　　　　　n. 校准　v. 调整；使成一行
discrimination [dɪˌskrɪmɪˈneɪʃən]　　n. 歧视；区别，辨别；识别力
rigidity [rɪˈdʒɪdəti]　　　　　　　　　n. 硬度，刚度

Notes 注释

1. high speed picking　高速搬运
2. vision sensor　视觉传感器
3. intelligent palletizing robot　智能码垛机器人
4. visual inspection　视觉分拣
5. arc welding robot　弧焊机器人
6. servo technology　伺服技术
7. acceleration performance　加速性能
8. drive mechanism　驱动机构
9. long arm type robot　长臂式机器人
10. mounting configuration　安装配置
11. pneumatic enclosure　气动外壳
12. impact force　冲击力
13. high pressure stream cleaning　高压喷流清洗
14. palletizing system　物流系统

Part 3 The Structure of FANUC Robot

Here we focus on FANUC's typical robot product LR Mate 200iD/4S robot, part of FANUC's mini six-axis industrial robots, as shown in Figure 7-8. FANUC LR Mate 200iD/4S robot is generally composed of three parts: manipulator, R-30iB Mate controller and iPendant.

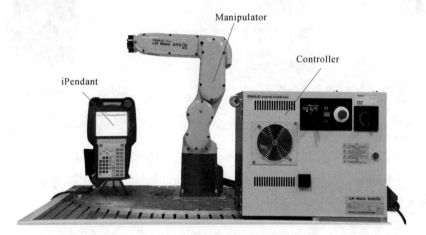

Figure 7-8 The Composition Structure of LR Mate 200iD/4S Robot

Manipulator

Manipulator is also called robot body. It is the main mechanical arm of industrial robots; the actuator used to complete the required tasks. It is mainly composed of robotic arm, driving device, transmission device and internal sensors. For a six-axis robot, the robotic arm mainly includes a base, a waist, two arms(upper arm and forearm) and a wrist, as shown in Figure 7-9.

Controller

The FANUC R-30iB Mate is a high-performance controller that brings a higher level of productivity to the production floor. It offers hardware and the latest advances in network communications, integrated iRVision, and motion control functions. Additionally, FANUC has minimized the controller space, saving floor space for the manufacturer, or allowing the manufacturer to stack controllers for multi-robotic installations.

This controller also helps to save energy as it requires less power consumption with its external power switch. The R-30iB Mate also has a cooling fan auto-stop that reduces power by switching off during breaks, a brake control function that reduces power through the automatic braking motor if the robot is idle for an elongated period of time, and a ROBOGUIDE power optimization function that minimizes power and optimizes energy savings for the consumer.

The controller also has an optional energy-saving design that recovers kinetic energy

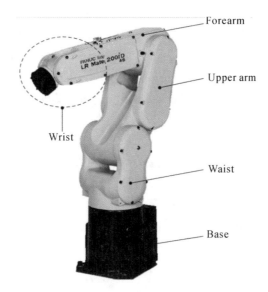

Figure 7-9 Robotic Arm of LR Mate 200iD/4S Robot

during braking and returns it to the system to be reused during the next cycle.

While the controller increases energy savings, and in turn, decreases energy costs, it also enhances the productivity of any company through an enhanced brake release, robot motion optimization and smart vibration control for the robot arm to reduce cycle time and produce smoother movement.

iPendant(Teach Pendant)

The iPendant is a hand held operator unit used to perform almost all of the tasks involved when operating a FANUC robot system.

The surface structure of the iPendant is shown in Figure 7-10.

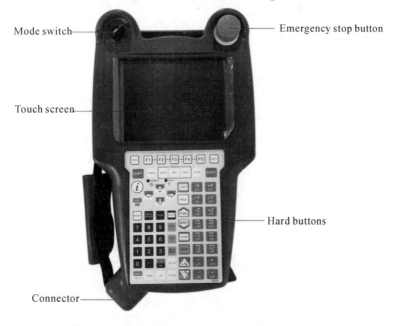

Figure 7-10 The Surface Structure of the iPendant

The FANUC iPendant is a highly customizable, portable display and operation panel. The iPendant has a touch screen with 4D graphics that displays process information and the actual process path, allowing for a much easier set up. If necessary, a touch panel interface can be added as an option and all the keys can be customized with the transparent key sheet. Also, the connection unit enables the iPendant to be removed for portability or security.

Vocabulary 词汇

controller [kənˈtrəʊlə]　　　　n. 控制器；管理员；主计长
portability [ˌpɔːtəˈbɪləti]　　　n. 可移植性
security [sɪˈkjʊərɪti]　　　　　n. 安全 adj. 安全的；保安的
customizable [ˈkʌstəmaɪzəbəl]　[计]可定制的

Notes 注释

1. iPendant　示教器（FANUC）
2. high-performance controller　高性能控制器
3. network communication　网络通信
4. stack controller　堆栈控制器
5. reduce power　降低功耗
6. switch off　关掉
7. energy saving　节能
8. motion optimization　动作优化
9. smart vibration control　智能振动控制
10. portable display　便携式显示器
11. operation panel　操作面板
12. touch screen　触摸屏
13. smart vibration　智能振动

Part 4 Typical Robot—LR Mate 200iD/4S

Here we focus on a typical robot product—LR Mate 200iD/4S robot, part of FANUC's mini six-axis industrial robots.

LR Mate 200iD/4S

LR Mate 200iD/4S is a human arm sized mini robot. The slim arm minimizes interference to peripheral devices at narrow space and the lightest mechanical unit in its class realizes easy integration into a machine or upside-down mounting on a frame. Additionally, high rigidity arm and the most advanced servo technology enable smooth motion without vibration in high speed operation. Wrist load capacity of LR Mate 200iD/4S is enhanced extremely and it makes efficiency to increase by handling plural work pieces. Because sensor cable, auxiliary axis cable, solenoid valve, air tube and I/O cable for device are integrated in the arm, it realizes easy hand cabling.

The features of LR Mate 200iD/4S robot are shown in Table 7-1 below.

Table 7-1 The Features of LR Mate 200iD/4S

Features	
Model	LR Mate 200iD/4S
Reach	550 mm
Max. Load Capacity at Wrist	4 kg
Installation	Floor, Upside-down, Angle mount
Repeatability	±0.02 mm
Mass	20 kg
Input Power Capacity	1.2 kVA

The movement of the LR Mate 200iD/4S robot is shown in Table 7-2 below.

Table 7-2 The Movement of LR Mate 200iD/4S

Movement		
Axis	Motion Range	Maximum Speed
J1 (Axis 1)	340°	460°/s
J2 (Axis 2)	230°	460°/s
J3 (Axis 3)	402°	520°/s
J4 (Axis 4)	380°	560°/s
J5 (Axis 5)	240°	560°/s
J6 (Axis 6)	720°	900°/s

HRG-HD1XKB

HRG-HD1XKB industrial robot skills assessment training platform (standard edition), as shown in Figure 7-11, is a universal six-axis robot training platform, can be installed with any brand of compact six-axis robot. Combined with a variety of automation mechanism, with standardized teaching modules for industrial applications, the training platform is mainly used for teaching and skills assessment of industrial robot technician.

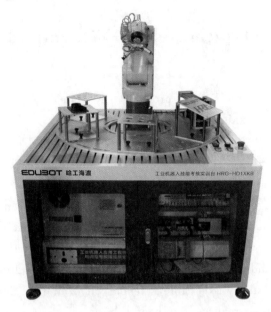

Figure 7-11　Industrial Robot Skills Assessment Training Platform(HRG-HD1XKB)

Asynchronous Conveyor Belt Module

After running conveyor belt, first, put the workpiece on the conveyor belt, the workpiece runs along the conveyor belt. Then, when the workpiece arrives at the end of the conveyor belt, where is installed with photoelectric switch, the switch detects the material and sends a feedback signal to the system, and the conveyor belt will stop. At last, the robot moves to the end of the conveyor belt, grabs and places the workpiece on the tray. The above demonstrates the flow processes of handling and storage in the production line, as shown in Figure 7-12.

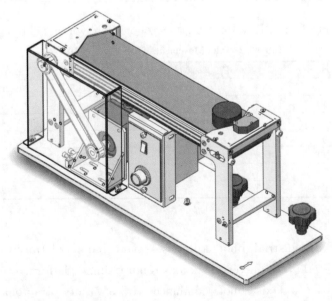

Figure 7-12　Asynchronous Conveyor Belt Module

Vocabulary 词汇

smooth [smuːð]   *adj.* 光滑的 *vt.* 使光滑 *n.* 平滑部分

Notes 注释

1. mini robot   迷你机器人
2. upside-down mounting   顶吊安装;吊装
3. sensor cable   传感器电缆
4. auxiliary axis cable   辅助轴电缆
5. solenoid valve   电磁阀
6. air tube   空气管
7. I/O cable   I/O 电缆
8. hand cabling   手工布线

Unit 8　SCARA Robot

Part 1　EPSON SCARA Robot

EPSON is the world's leading maker of SCARA robots. It began developing robots in 1981. Over the past three decades, EPSON has used its robot expertise to produce many successful generations of EPSON robots known for outstanding precision, speed and reliability.

The following is a brief description of the main series of EPSON SCARA robots (the specific parameters of the specifications are from EPSON officially announced data).

G Series

The space-saving robots in the G series are designed with EPSON's unique Smart Motion Control technology and provide among the industry's best speed and accuracy, as shown in Figure 8-1.

EPSON enhanced the specifications and functions important for users to design and develop the system and enabled system to be optimized for multi-product production, some are supporting environment with water and oily mist.

The following is the main features of G series of EPSON SCARA robots.

(1) Quick, accurate movements shorten work time.

(2) Precise and smooth straight and arch motions improve the quality of sophisticated work.

(3) A ductless design with inner wiring increases the motion range by 20%.

(4) Mounting options enable robots to be installed on tabletop, wall, or ceiling, for the best layout and most space-efficient system.

RS Series

EPSON RS series robots are more than just SCARA robots. We call them SCARA+, as shown in Figure 8-2. Unlike the normal SCARA robots, the EPSON RS series have a unique ability to move the second axis underneath the first axis thus allowing them to move throughout the entire workspace. Normal SCARA robots have to move around themselves thus leaving a big hole in the middle they cannot reach. Thus EPSON RS SCARA+ robots have both work envelope and speed advantage. EPSON RS series SCARA+ robots have repeatability down to 0.010 mm and cycle time of 0.339 s.

RS series are literally zero footprint robots. They are capable of easy integration into compact assembly cells. The unique work space design of RS series robots improves cycle time compared with other "old" SCARA designs. This means more parts processed in less time, while using a fraction of floor space which results in more profits.

Figure 8-1 G series Robot

Figure 8-2 RS series Robot

LS Series

LS series robots were created as the reduced cost solution for factories looking for maximum value without giving up performance, as shown in Figure 8-3. But as with all EPSON robots, they are still packed with the same performance and reliability. In addition, EPSON LS series robots are able to meet the demanding space requirements in factories today by providing a small footprint solution with superior workspace usage. With outstanding cycle time, industry leading ease of use and reliability, they are the robots of choice for high speed applications where superior performance is required at a value price.

T Series

T series compact SCARA robots feature a built-in controller that eliminates the need to deal with complicated cabling during setup and maintenance, and a motor unit without battery that improves cost efficiency to help keep total operating cost low, as shown in Figure 8-4.

T series provides an I/O communication port closer to the end effector. This port makes it easier to connect cables to the end effector and supply power to it. There is no longer any need to route a long cable to the controller. The cable conduit, which contains pneumatic hoses as well as electrical cables, is shorter than in previous models. The shorter design gives it stability while the robot is moving, making it easier to route cables outside the conduit.

Figure 8-3 LS series Robot

Figure 8-4 T series Robot

Vocabulary 词汇

quick [kwɪk]   n. 核心 adj. 快的;迅速的 adv. 迅速地
accurate [ˈækjʊrət]   adj. 精确的
precise [prɪˈsaɪs]   adj. 精确的;明确的;严格的
ductless [ˈdʌktlɪs]   adj. 无导管的
performance [pəˈfɔːməns]   n. 性能;绩效;执行;表现
repeatability [rɪˌpiːtəˈbɪlɪti]   n. 重复性;[计]可重复性;再现性
reliability [rɪˌlaɪəˈbɪləti]   n. 可靠性

Notes 注释

1. EPSON robot   爱普生机器人
2. space-saving robot   节省空间的机器人
3. smooth straight   平滑直线
4. arch motion   圆弧运动
5. built-in controller   内置控制器
6. cable conduit   电缆导管
7. pneumatic hose   气动软管

Part 2　YAMAHA SCARA Robot

YAMAHA is one of the world's leading makers of SCARA robots. The first YAMAHA robot was SCARA robot. Since the first SCARA robot called "CAME" was produced in 1979, about 30 years of SCARA robot innovations have continually appeared.

YK series of YAMAHA SCARA robots, which are the top of rich product series in industry with a totally beltless structure, play out the limit of the level characteristics of multi-joint robots. The following is a brief description of the main models of YK series(the specific parameters of the specifications are from YAMAHA officially announced data).

YK350TW

YK350TW is the omnidirectional type of SCARA robot, which has 350 mm maximum reach, as shown in Figure 8-5. It resolves the shortcomings of previous SCARA and parallel-link robots and offers both superior positioning accuracy and high speed. Featuring a ceiling-mount configuration with a wide arm rotation angle, the YK350TW can access any point within the full 1 000 mm downward range.

YK150XG

YK150XG is a tiny SCARA robot, which has a 1 kg payload capacity, as shown in Figure 8-6. Only this robot has a completely beltless structure in its class. The robot is super-compact, but achieves the overwhelming high rigidity and high accuracy. A totally beltless structure is achieved by using a ZR axis direct coupling structure. This direct drive structure drastically reduces wasted motion. Additionally, a drastic improvement in maximum speed of YK150XG is made by changing the gear ratio and maximum motor rpm. The cover on the YK150XG, which can be removed from the front or upwards, is separate from the cable so maintenance tasks are easy.

Figure 8-5　YK350TW Robot　　　　Figure 8-6　YK150XG Robot

YK350XG

The small size SCARA robot YK350XG with payload capacity of 5 kg is suitable for assembly, transfer, pushing work, and sealing of small components, as shown in Figure 8-7. It has a speed reducer directly coupled to the tip of the rotating axis. The R axis produces an extremely high allowable inertia moment which delivers high speed operation compared to structures where positioning is usually done by a belt after decelerating. Changing the cable layout makes the overall cable height lower than the unit cover. Also, utilizing a motor with a small overall length and extrusion material base yields the smallest dimensions among equipment in the same class.

YK550XG

YK550XG, a medium type SCARA robot with 500 mm maximum reach, has high accuracy, high speed, excellent maintenance ability and amazing tolerable moment of inertia, as shown in Figure 8-8. It however can automatically set an optimal maximum acceleration and deceleration using the arm status of starting operation and ending operation. This capability means that just entering the initial payload will prevent the robot from exceeding tolerance values for motor peak torque and speed reducer allowable peak torque. So full power can be extracted from the motor whenever needed and a high level of acceleration or deceleration is maintained.

Figure 8-7　YK350XG Robot

Figure 8-8　YK550XG Robot

YK800XG

Large type SCARA robot YK800XG is suitable for assembly and transfer of large objects or heavy objects, as shown in Figure 8-9. The position detector of YK800XG is a resolver. The resolver has a simple yet strong structure using no electronic components or elements and so it has great features such as being extremely tough in harsh environment as well as a low breakdown rate. The resolver structure has none of the detection problems that occur in other detectors such as optical encoders whose electronic components breakdown or suffer from moisture or oil that sticks to the disk. Moreover, mechanical specifications for both absolute and incremental specifications are common to all controllers so the robot can switch to either absolute or incremental specifications just by setting parameters. Also even if the absolute battery is completely worn down, the SCARA can operate on incremental specifications so in the unlikely event of trouble users can feel secure knowing that there will be no need to stop the production line.

YK400XR

New SCARA robot YK400XR is developed by redesigning even mechanical components of conventional models. This SCARA robot gives the high quality and high performance, and yet superior cost performance. The robot with 400 mm arm length and payload capacity of 5 kg achieves the standard cycle time of 0.45 s, as shown in Figure 8-10. Combined with the new controller RCX340, high function control can be achieved.

Figure 8-9　YK800XG Robot

Figure 8-10　YK400XR Robot

Vocabulary 词汇

omnidirectional [ˌɒmnɪdəˈrekʃənəl]　　　　　*adj.* 全方向的
produce [prəˈdjuːs]　　　　　　　　　　　　*n.* 产品
resolver [rɪˈzɒlvə]　　　　　　　　　　　　*n.* 溶剂；[电子]分解器；下决心者

Notes 注释

1. SCARA robot　SCARA 机器人
2. beltless structure　无带结构
3. tiny SCARA Robot　小 SCARA 机器人
4. direct coupling　直接耦合
5. pushing work　推动工作
6. inertia moment　惯性力矩
7. high accuracy　高准确率
8. excellent maintenance ability　良好的维护能力
9. heavy object　重物
10. optical encoder　光学编码器
11. electronic components breakdown　电子元件故障
12. suffer from moisture　受潮
13. oil that sticks to the disk　粘在磁盘上的油
14. high quality　高质量
15. high performance　高性能
16. superior cost performance　卓越的性价比
17. high function control　高功能控制

Part 3　The Structure of SCARA Robot

SCARA robot is generally composed of three parts: manipulator, controller and teach pendant, as shown in Figure 8-11.

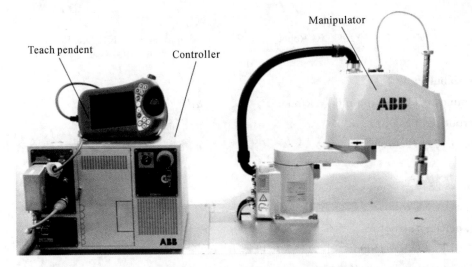

Figure 8-11　The Composition Structure of SCARA Robot

Manipulator

Manipulator is also called robot body. It is the main mechanical arm of industrial robots; the actuator used to complete the required tasks. It is mainly composed of robotic arm, driving device, transmission device and internal sensors. For a SCARA robot, the robotic arm mainly includes a base, upper arm and forearm, as shown in Figure 8-12.

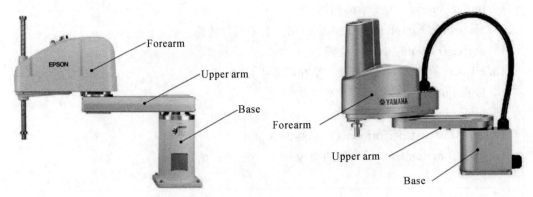

Figure 8-12　Robotic Arm of SCARA Robot

Controller

Here we focus on the RC90 controller of EPSON SCARA robot and the RCX340 controller of YAMAHA SCARA robot.

1. RC90 Controller

The EPSON RC90 controller is a powerful yet inexpensive controller used for the EPSON LS series SCARA Robots, as shown in Figure 8-13. Like other EPSON controller products, RC90 provides the ultimate experience in ease of use and, at a nicely reduced price. RC90 can control EPSON LS3 and LS6 SCARA robots and provides superior PowerDrive servo control for super smooth motion, fast accel/decel time, and fast cycle time. EPSON RC90 controller has high performance at a good price.

Figure 8-13 The RC90 Controller of EPSON SCARA Robot

The RC90 comes standard with EPSON RC + Controls software and lots of fully integrated options. As the industry leader in ease of use, EPSON RC+ wizards, point and click setup, jog and teach window, integrated debugger, Epson Smartsense and many other features help reduce overall development time as compared to competitive robot systems. In addition, options such as Vision Guidance,. NET support, GUI Builder, DeviceNet, Profibus and many more are fully integrated into the EPSON RC + development environment to maximize performance and ease of use.

2. RCX340 Controller

The third generation model of RCX controller RCX340 is now put on the market by reviewing all functions to further improve the functions of the RCX controller, as shown in Figure 8-14.

Figure 8-14 The RCX340 Controller of YAMAHA SCARA Robot

The RCX340 controller allows high-speed communication among the controllers. As the master controller sends operation commands to each slave controller, programs and points can be controlled only with the host master controller. Additionally, as the RCX340 controller flexibly supports multi tasks, the use of the PLC can be simplified. Simultaneous start and arrival of each robot can also be controlled freely. Complicated and precise robot system using many axes can be constructed more simply at low costs. As a new servo motion engine is incorporated into the controller, various operations can be linked. Newly developed algorithms achieve reduction of the positioning time and improvement of the tracking accuracy. The volume ratio is reduced approx. 85% when compared to the conventional four-axis controller to achieve the compact design and make the installation inside the control panel easy. RS 232C and Ethernet ports are provided as standard equipment in the RCX controller. A wide variety of high-speed and large capacity field networks, such as CC-Link, DeviceNet^{TM}, and EtherNet/IP^{TM} are supported as options. Connections with general purpose servo amplifier or other company's VISION are easy. Therefore, the RCX340 is called "connectable controller".

Teach Pendant

Teach pendant, the human-computer interaction interface, can be operated to move by the operator. For SCARA, however, some manufacturers are using teach pendant as an optional accessory, such as EPSON and YAMAHA, which can be used to control the robot through the corresponding software. So in this part there is no introduction about teach pendant of SCARA robot.

Vocabulary 词汇

wizard ['wɪzəd]   *n.* 术士；巫师；奇才
jog [dʒɒg]   *vt.* 轻推；蹒跚行进 *n.* 慢跑；轻撞

Notes 注释

1. ease of use   使用方便
2. point and click set up   点击设置
3. teach window   示教窗口
4. itegrated debugger   集成调试器
5. high-speed communication   高速通讯
6. connectable controller   可连接控制器

Part 4 Applications of SCARA Robot

At present, SCARA robots are ideally suited for the medical, automotive, electronics, food, lab automation, semiconductor, plastic, appliance, aerospace industries and many more. They can be used for a wide variety of applications, such as parts handling and assembly.

1. Robotic Handling

SCARA robots are characterized by the combination of two poles similar to man's arm, which can trend into assignments and then back in limited space. Thus the robots are suitable for handling and picking up, such as integrated circuit boards, as shown in Figure 8-15.

2. Robotic Assembly

SCARA robot is elective compliance in the X and Y direction, but rigid in the Z direction, which is especially suitable for assembly work. SCARA robot was the first to be used to assemble printed circuit boards and electronic components, as shown in Figure 8-16.

Figure 8-15 Handling of SCARA Robots   Figure 8-16 Assembly of SCARA Robots

Industrial Robot Skills Assessment Training Platform(HRG-HD1XKS)

HRG-HD1XKS SCARA robot skills assessment training platform, as shown in Figure 8-17, is a universal SCARA robot training platform. The platform is cored on SCARA robots commonly used in industries, with a variety of standardized teaching modules for industrial applications. It can be used for the teaching of industrial robot virtual simulation, PLC, programming, etc. The platform can be used to learn the robot vision, object assembly, handling, palletizing and other practical functions, and is mainly used for teaching and skills assessment of industrial robot technician.

Handling Module

The module can be used to learn handling applications of robot, the workpiece or

Figure 8-17 Industrial Robot Skills Assessment Training Platform(HRG-HD1XKS)

material is transferred from one station of the tray to another. It can be used to learn programming skills of robot motion, such as linear motion, circular motion, curve movement, as shown in Figure 8-18.

Figure 8-18 Handling Module

Vocabulary 词汇

medical ['medɪkəl]	*adj.* 医学的；药的 *n.* 医生；体格检查
automotive [ˌɔːtə'məʊtɪv]	*adj.* 汽车的；自动的
electronics [ɪˌlek'trɒnɪks]	*n.* 电子学；电子工业
food [fuːd]	*n.* 食物；养料
semiconductor [ˌsemikən'dʌktə]	*n.* [电子][物]半导体
appliance [ə'plaɪəns]	*n.* 器具；器械；装置
compliance [kəm'plaɪəns]	*n.* 顺从；服从；承诺

Notes 注释

1. aerospace industry 航空航天工业
2. lab automation 实验室自动化
3. integrated circuit 集成电路
4. electronic component 电子元器件

Unit 9　Industry Application of Robot

Part 1　Painting Robot

If your company has a painting application, why not automate it with a robot? Painting automation is a simpler, safer and better method than any manual painting process. Furthermore, industrial painting robots are much more accessible than they used to be. Not only are there more models on the market, but also they are more affordable than ever before.

Then what would be like if all go with automation? It is about the cost of doing business and how to decrease that cost. Well, if you're painting, the way to decrease costs across the board is to automate!

The Cost of Time

The expression "time is money" definitely applies to painting and coating jobs. Manual painting costs more because it takes longer. Not only does a worker operate more slowly than a painting robot, but workers also have to be allowed breaks, lunches and vacations. The repetitive nature of the painting task can also cause fatigue, stress and injury. Robots, on the other hand, are capable of painting 24 hours a day, 365 days a year. They work efficiently no matter how long they have been running, increasing throughput, while never decreasing quality.

The Cost of Materials

When a worker is painting and coating manually, mistakes like overspray can waste materials and decrease the quality of products, possibly damaging them. The quality of manual painting is never consistent. Overall, it is a messy process that will end up costing your company more money.

On the other hand, robots conserve paint and work with incredible precision and consistency. Typical paint savings when using robotic automation is 15～30 percent. Since they are programmed to spray the same amount of materials on every product, they have fewer overspray problems and create a consistent coating on each part every time.

The Cost of Safety

Paint contains hazardous materials like Xylene and Toluene. These substances are common in several different paint and coating materials, and it is very dangerous for humans to work around them for long period of time because of the toxic fumes the chemicals produce. Instead, you can choose to automate with painting robots, which control and isolate the above hazards. Workers are removed from the dangers and placed in supervisory roles.

What is a painting robot?

"Painting robot" is an industry term for a robot that has two major differences from all other standard industrial robots.

(1) Explosion proof arms. Painting robots are built with explosion proof robot arms, meaning that they are manufactured in such way that they can safely spray coatings that create combustible gasses. Usually these coatings are solvent based paints, when applied, it must create an environment that must be monitored for fire safety.

(2) Self-contained paint systems. When painting robots were first designed, they only had one function—to work safely in a volatile environment. As acceptance and use expanded, painting robots grew into a unique subset of industrial robots, not just traditional robots with explosion proof options. Painting robots now have the ability to control all aspects of spray parameters. Fan air, atomization air, fluid flow, voltage, etc, can all be controlled by the robot control system.

The ABB painting robot is shown in Figure 9-1.

Figure 9-1 ABB Painting Robot

Vocabulary 词汇

manual ['mænjuəl]	adj. 手工的;体力的 n. 手册,指南
model ['mɒdl]	n. 模型;典型;模范;模特儿;样式
automate ['ɔːtəmeɪt]	vt. 使自动化 vi. 自动化

fatigue [fə'ti:g]　　　　　　　　n. 疲劳,疲乏 vt. 使疲劳 vi. 疲劳 adj. 疲劳的
throughput ['θru:pʊt]　　　　　 n. 生产量,生产能力
overspray ['əʊvəspreɪ]　　　　　n. 超范围的喷涂 vt. 过喷
consistency [kən'sɪstənsi]　　　n. [计]一致性;稠度;相容性
spray [spreɪ]　　　　　　　　　n. 喷雾;喷雾器 vt. 喷射 vi. 喷
substance ['sʌbstəns]　　　　　n. 物质;实质;资产;主旨
isolate ['aɪsəleɪt]　　　　　　　vt. 使隔离;使孤立 adj. 隔离的
supervisory ['sju:pəˌvaɪzəri]　　adj. 监督的
solvent ['sɒlvənt]　　　　　　　adj. 有偿付能力的;有溶解力的 n. 溶剂;解决方法
monitor ['mɒnɪtə]　　　　　　　v. 监督;监控,监听;测定
volatile ['vɒlətaɪl]　　　　　　adj. [化学]挥发性的;不稳定的 n. 挥发物;有翅的动物
subset ['sʌbset]　　　　　　　　n. [数]子集;子设备;小团体
parameter [pə'ræmɪtə]　　　　　n. 系数;参量
voltage ['vəʊltɪdʒ]　　　　　　n. [电]电压

Notes 注释

1. painting robot　喷涂机器人
2. painting application　喷涂应用
3. painting automation　喷涂自动化
4. manual painting　手工喷涂
5. industry term　行业术语
6. explosion proof arm　防爆手臂
7. combustible gass　可燃气体
8. self-contained paint system　独立的喷涂系统
9. atomization air　雾化空气
10. fluid flow　流体流动

Part 2 Welding Robot

The welding robot is an industrial robot engaged in welding. Industrial robot is a versatile, reproducible manipulator with three or more programmable axes for industrial automation. In order to adapt itself to different uses, the mechanical interface of the robot's last axis is usually a connecting flange, which can be attached to different tools or end effectors. The welding robot is an industrial robot of which a welding tong or torch is in the end, so that it can weld, cut or hot spray.

With the development of electronic technology, computer technology, numerical control and robot technology, since the beginning of the 1960s, automatic welding robot began to be used for production; its technology has become increasingly mature. It has mainly the following advantages:

(1) to stabilize and improve the quality of welding, welding quality can be reflected in the form of value;

(2) to improve labor productivity;

(3) to improve the labor intensity of workers, can work in a harmful environment;

(4) to reduce requirements for workers' operating techniques;

(5) to shorten the product replacement of the preparation cycle, reduce the corresponding equipment investment.

Therefore, welding robot has been widely used.

Welding robot mainly includes two parts—robot and welding equipment. The robot consists of the robot body and the control cabinet (hardware and software). Take arc welding as an example, the welding equipment consists of the welding power (including its control system), wire feeder, welding gun and other components. For intelligent robot it should also have a sensing system, such as laser or camera sensors and their control devices.

In addition, if the workpiece in the entire welding process is without displacement, you can use the fixture to locate the workpiece on the table; this system is the most simple. However, in the actual production, many workpieces in the welding need to change position, so that the weld in a better position (posture) is welded. For this situation, the positioning machine and the robot can separately move, that is the robot welds again after the displacement machine changing position; and they can also be moving at the same time, the displacement machine is changing position while the robot is welding, that is often said that the displacement machine and robot coordinate to move. At this time the movement of the positioning machine and the movement of the robot are compound movement, so that the movement of the welding gun relative to the workpiece can meet the weld trajectory, and can meet the requirements of the welding speed and the welding gun's posture. In fact, the shaft of the positioner has become part of the robot, and this welding robot system can be up to 7~20 axes or more.

Vocabulary 词汇

torch [tɔːtʃ]	n. 火把;火炬 vi. 像火炬一样燃烧
stabilize [ˈsteɪbɪlaɪz]	vt. 使稳固,使安定 vi. 稳定,安定
displacement [dɪsˈpleɪsmənt]	n. 取代;移位
component [kəmˈpəʊnənt]	n. 部件;组件;成分
device [dɪˈvaɪs]	n. 装置;策略;图案
fixture [ˈfɪkstʃə]	n. 设备;固定装置
workpiece [ˈwɜːkpiːs]	n. 工件;轧件;工件壁厚
posture [ˈpɒstʃə]	n. 姿势;态度 vi. 摆姿势
separately [ˈsepərɪtli]	adv. 分别地;分离地;个别地
coordinate [kəʊˈɔːdɪnɪt]	adj. 并列的;同等的 vt. 调整;整合
shaft [ʃæft]	n. [机]轴;箭杆 vt. 利用

Notes 注释

1. end effector　末端执行器
2. welding tong　焊钳
3. hot spray　热喷雾
4. electronic technology　电子技术
5. numerical control　数控
6. labor productivity　劳动生产率
7. labor intensity　劳动强度
8. preparation cycle　准备周期
9. welding equipment　焊接设备
10. arc welding　电弧焊
11. wire feeder　送丝机;送丝装置
12. welding gun　焊枪
13. intelligent robot　智能机器人
14. sensing system　传感系统
15. camera sensor　相机传感器
16. positioning machine　定位机
17. displacement machine　变位机
18. compound movement　复合运动
19. weld trajectory　焊接轨迹

Part 3　Handling Robot

Handling robots are industrial robots that can be automated for handling operations. The earliest handling robots appeared in 1960 in the United States, the two styles of Versatran and Unimate robots were used for handling operations for the first time. Handling operation is that uses a device to hold the workpiece, and moves it from one process place to another. The handling robot can be fitted with different end actuators to perform various workpieces handling in different shapes and conditions, that greatly reduces the heavy manual labor of human. At present, more than 100 000 handling robots are used in automatic assembly line, palletizing, container and so on. Some developed countries have drawn up the maximum amount of manual handling, the work which is more than the limit must be carried by the handling robot.

The handling robot is a high-tech in the field of modern automatic control, which involves the academic areas of mechanics, mechanology, hydraulic-air pressure technology, automatic control technology, sensor technology, single chip microcomputer technology and computer technology, and it has become an important component of modern machinery manufacturing system. Its advantage is that you can program to complete a variety of tasks, they have advantages in their own structures and performances which people and the machines own. In particular, they reflect the artificial intelligence and adaptability.

There are tandem joint robots, horizontal joint robots(SCARA robots), Delta parallel joint robots and AGV handling robots in common. Above all kinds of robots in the handling industry have their own characteristics.

The tandem joint robot usually has four or six rotating axes, and is similar to the human arm, more flexible, the load of them is larger; they are used for handling heavier items.

SCARA robot is a horizontal multi-joint robot, and has faster work frequency, that is particularly suitable for handling goods in laboratory automation, medicine, consumer electronics, food, automotive, PC peripherals, semiconductors, plastics, home appliances and aerospace industry.

Delta robot belongs to parallel robot. Delta robots are mainly used for handling, assembling food, medicine and electronic products and so on. Delta robot is widely used in the market because of its light weight, small size, fast moving speed, accurate positioning, low cost and high efficiency.

China is in the critical moment of the industrial transformation and upgrading, more and more enterprises introduce industrial robots in the manufacturing process, it will profoundly affect all aspects of Chinese-made.

Vocabulary 词汇

container [kənˈteɪnə]   n. 集装箱；容器
adaptability [əˌdæptəˈbɪləti]   n. 适应性；可变性；适合性
horizontal [ˈhɒrɪzɒntl]   adj. 水平的；地平线的；同一阶层的 n. 水平线；水平面
parallel [ˈpærəlel]   n. 平行线 vt. 使……与……平行 adj. 平行的；类似的
rotate [ˈrəʊteɪt]   vi. 旋转；循环
plastic [ˈplæstɪk]   n. 塑料；整形外科；外科修补术
guide [gaɪd]   vt. 引导；带领；操纵
upgrading [ˌʌpˈgreɪd]   vt. 使升级；提升；改良品种

Notes 注释

1. automatic assembly line   自动装配线
2. hydraulic-air pressure technology   液压气压技术
3. single chip microcomputer technology   单片机技术
4. artificial intelligence   人工智能
5. tandem joint robot   串联关节机器人
6. horizontal joint robot   水平关节机器人
7. Delta parallel joint robot   Delta 并联关节机器人
8. work frequency   工作频率
9. consumer electronics   消费类电子产品
10. home appliance   家用电器
11. aerospace industry   航空航天工业
12. accurate positioning   准确定位
13. automatic logistics transport   自动物流运输

Part 4 Assembly Robot

"Assembly robot" is an industry term for a robot that is used in lines of industrial automation production to assemble parts or components and it is the core of the flexible assembly system automation equipment. It's used for lean industrial processes and has expanded production capabilities in the manufacturing world. The assembly line robot can increase production speed and consistency. The tools at the end of arm can be customized for each assembly robot to cater to the manufacturing requirements and used for all applications.

Assembly operations with industrial robots can make the process faster, improve efficiency and accuracy.

1. Robotic Assembly Offers Many Benefits

In assembly automation, robots may be equipped with vision technology to accommodate inconsistently located parts or components. In addition, robotic vision is also able to improve efficiency and accuracy. Robots can perform tedious and dull assembly tasks leaving factory members to do other jobs, whilst at the same time quality is improved.

2. Round the Clock Production Is a Cost Effective Option

Robots will work 24 hours a day all year round without the need for a break. They will eliminate downtime, reduce labour costs and provide a high return on investment.

Assembly robots are widely used in all kinds of electrical appliance manufacturing, auto and its components, computer, medical, food, solar, toys, machinery and electronic products and their components and other industries, as shown in Figure 9-2.

Classification of Assembly Robot

Currently, there are four types of assembly robots in industrial applications that are common: Cartesian coordinate robots, six-axis articulated robots, SCARA robots and Delta parallel robots.

The Cartesian coordinate robot is mainly used in the assembly of energy-saving lamps, electronic products and LCD panels. The six-axis robot has the widest range of applications and can accommodate most situations. In addition, SCARA robot is widely used in the assembly of products such as electronics, machinery and light industry and the Delta parallel robot is mainly used in the field of IT and electronic assembly.

The Structure of Assembly Robot's System

The assembly robot's system consists mainly of manipulator, controller, teach pendant, assembly operating system and peripheral equipment, as shown in Figure 9-3.

1. Assemble Operating System

The assembly operating system consists mainly of the transport-type end effector and vacuum negative pressure station, and the manipulator has the visual system.

igure 9-2 Application of Industrial Robot in Assembly

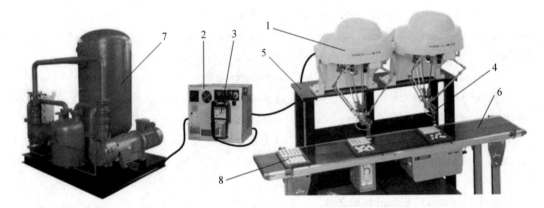

Figure 9-3 The Structure of Assembly Robot's System
1—Manipulator;2—Controller;3—Teach pendant;4—End effector;5—Platform;6—Conveyor;
7—Vacuum negative pressure station;8—Workpiece

2. Peripheral Equipment

The peripheral equipment of assembly robot's system, such as fence, the robot platform, transmission device, the device to put workpiece, parts feeder, facilitates the system to complete the whole assembly.

Vocabulary 词汇

customize [ˈkʌstəmaɪz]	vt. 定制,定做;按规格改制
accuracy [ˈækjʊrəsi]	n. [数]精确度,准确性
tedious [ˈtiːdiəs]	adj. 沉闷的;冗长乏味的
solar [ˈsəʊlə]	adj. 太阳的;日光的 n. 日光浴室
eliminate [ɪˈlɪmɪneɪt]	vt. 消除;排除

downtime ['daʊntaɪm]   n. [电子]故障停机时间
fence ['fens]   n. 防护物 vt. 防护；用篱笆围住

Notes 注释

1. assembly robot   装配机器人
2. vision technology   视觉技术
3. six-axis articulated robot   六轴铰链机器人
4. parallel robot   并联机器人
5. energy-saving lamp   节能灯
6. light industry   轻工业
7. operating system   操作系统
8. vacuum negative pressure station   真空负压站
9. the robot platform   机器人平台
10. transmission device   传输设备
11. the device to put workpiece   工件摆放装置
12. parts feeder   上料机

Part 5　Polishing Robot

"Polishing robot" is an industry term for robots that can polish automatically, it is widely used in the fields of 3C, sanitary ware, IT, auto parts, industrial parts, medical devices and civil products, as shown in Figure 9-4.

Figure 9-4　Application of Polishing Robot in Auto Industry

Robotic polishing is the process of refining surface until they are smooth and shiny. This application is repetitive and tedious while requiring extreme consistency. Polishing robots are programmed to apply the appropriate pressure and move precisely in the right direction, for consistent, thorough, high-quality products.

1. The Advantages of Robotic Material Removal

With the flexibility, repeatability and extreme precision, polishing robots are possible to grind, trim or polish almost any material to achieve a consistent high-quality finish. These robots also improve production time while reducing waste. Robots save workers from both the drudgery and safety hazards associated with polishing. Polishing robots are unharmed by fumes and dust. Besides, robotic polishing is better for the environment because dry abrasive wheels are used instead of chemical solutions.

2. Methods for Handling the Process with a Robot

Currently, polishing robots are mostly the six-axis robots in industrial applications. Based on the different properties of the end effector, there are two main approaches for the robot—to hold the workpiece or the tool, as shown in Figure 9-5.

The polishing robot holding the workpiece is usually used to polish the relatively small workpiece. It grabs the workpiece without being polished by its end effector and makes this workpiece polished on the polishing machine. In addition, it is possible to add value to the system by letting the robot unload the finished part onto a conveyor or similar equipment. There is usually one or several tools around the polishing robot. However, robot holding

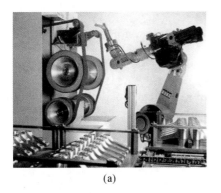 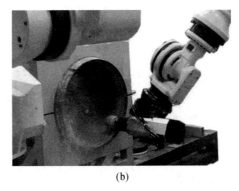

Figure 9-5 Classifications of Polishing Robot
(a) Holding the Workpiece   (b) Holding the Tool

the tool generally applies to large parts or workpieces that are heavy for the polishing robot. The handling of the workpiece can be done manually, and the robot automatically changes the required polishing tools from the tool rack. Usually applying the force control device in this system to ensure that the polishing pressure between the tool and the workpiece is consistent and to compensate the consumption of polishing head. The polishing quality can be uniformed by the force control device while the teach also can be simplified.

In practical application, it is also possible to have several robots working together for ultimate flexibility. One robot holds the part, while others manipulate the tool.

The Structure of Polishing Robot's System

The polishing robot's system of holding the tool consists mainly of manipulator, controller, teach pendant, the operating system of polishing and peripheral equipment, as shown in Figure 9-6.

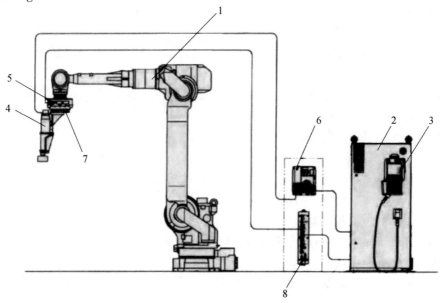

Figure 9-6 The Structure of Polishing Robot's System
—Manipulator; 2—Controller; 3—Teach Pendant; 4—End Effector; 5—Force Sensor; 6—Inverter;
7—Automatic Tool Changer(ATC); 8—Force Sensor Controller

1. The Operating System of Polishing

This system includes polishing power head, inverter, force sensor, force sensor controller and automatic tool changer(ATC).

2. Peripheral Equipment

The peripheral equipment of polishing robot's system, including fence, the robot platform, transmission device, the device to put workpiece, quieter, facilitates the system to complete the whole polishing.

Vocabulary 词汇

civil ['sɪvəl]	adj. 公民的；民间的；文职的
grind [ɡraɪnd]	vt. 磨碎；磨快 vi. 磨碎；折磨
trim [trɪm]	vt. 修剪；整理；装点 vi. 削减
drudgery ['drʌdʒəri]	n. 苦工，苦差事
dust [dʌst]	n. 灰尘；尘埃；尘土
hold [həʊld]	vt. 持有；保存 vi. 支持；有效
conveyor [kən'veɪə]	n. 输送机，[机]传送机
compensate ['kɒmpenseɪt]	vi. 赔偿；抵消 vt. 补偿，赔偿
consumption [kən'sʌmpʃən]	n. 消费；消耗；肺痨
manipulate [mə'nɪpjʊleɪt]	vt. 操纵；巧妙地处理；篡改
inverter [ɪn'vɜːtə]	n. 换流器；[电子]反相器

Notes 注释

1. polishing robot　打磨机器人

2. sanitary ware　洁具

3. auto part　汽车零件

4. refining surface　精制面

5. extreme precision　极度精准；非常精准

6. safety hazard　安全隐患

7. abrasive wheel　砂轮

8. chemical solution　化学溶液

9. tool rack　工具架

10. the operating system of polishing　打磨操作系统

11. polishing power head　打磨头

12. force sensor　力传感器

13. force sensor controller　力传感器控制器

14. transmission device　输送装置

15. automatic tool changer　自动换刀

Unit 10　New Types of Robots

Part 1　YUMI

The new era of robotic co-workers is here. YuMi is the result of years of research and development, making collaboration between humans and robots a reality, but there is also more significance.

ABB has developed a collaborative, dual-arm, small parts assembly robot solution that includes flexible hands, parts feeding systems, camera-based part location and state-of-the-art robot control. YuMi is a vision of the future. YuMi will change the way we think about assembly automation. YuMi is "you and me", working together to create endless possibilities.

Introduction of Structure

The IRB 14000 is ABB Robotics's first generation dual-arm robot with seven-axis each arm, it is designed specifically for manufacturing industries that use flexible robot-based automation, eg, 3C industry, as shown in Figure 10-1. The robot has an open structure that is especially adapted for flexible use, and can communicate extensively with external systems.

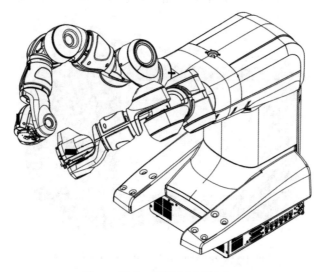

Figure 10-1　ABB YuMi Robot

Operating System

The robot is equipped with the controller (located inside the body of the robot) and robot control software, RobotWare. RobotWare supports every aspect of the robot system, such as motion control, development and execution of application programs, communication etc.

Additional Functionality

For additional functionality, the robot can be equipped with optional software for application support, for example communication features and advanced functions such as multitasking, sensor control etc.

Arm Axes

Each arm of YuMi has 7 axes, the multi degrees of freedom makes it as flexible as human hand, as shown in Figure 10-2.

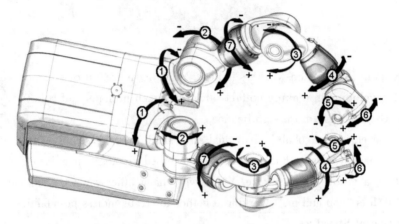

Figure 10-2　The Axes of YuMi

The arm configuration applies to both arms.

Safety Function

The safety functions are inherent design measures in the control system, contributing to power and force limiting. The teach pendant of ABB robot is shown in Figure 10-3.

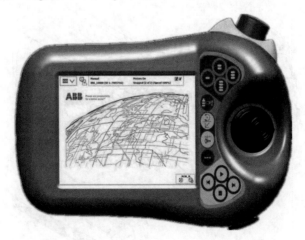

Figure 10-3　Teach Pendant of ABB Robot

Vocabulary 词汇

co-worker [ˈkəʊˌwɜːkə]　　　　　　　　　　　　n. 同事；合作者

collaborative [kəˈlæbərətɪv]   adj. 合作的；协作的
multitasking [ˈmʌltiˌtɑːskɪŋ]   n. 多重任务处理；多重任务执行
sensor [ˈsensə]   n. 传感器
power [ˈpaʊə]   n. 力量；电力 vt. 激励
force [fɔːs]   n. 力量；武力；军队；魄力
communication [kəˌmjuːnɪˈkeɪʃən]   n. 通讯；通信；交流

Notes 注释

1. flexible hand　灵巧手
2. parts feeding system　零件供料系统
3. manufacturing industry　制造业
4. operating system　操作系统
5. application program　应用程序
6. communication feature　通讯功能
7. multi degrees of freedom　多自由度
8. arm configuration　手臂配置

Part 2 SDA10F

The SDA10F is a dual-arm, fifteen-axis robot with incredible dexterity, freedom of movement in a compact footprint, as shown in Figure 10-4. Both arms can work together dramatically simplifying end-of-arm tools. Designed with patented servo actuators, all cables are routed through the arms. The FS100 is a powerful controller with unmatched open software architecture.

Figure 10-4 SDA10A Robot of Yaskawa

Key Benefits

Having the dexterity to perform complex tasks, dual seven-axis arms can work together or independently. Its slim design optimizes space; it provides "human-like" flexibility and range of motion, even in tight space. Simplified tooling reduces cost. It can be used in environments that are hazardous to humans. Labor savings justifies capital investment.

It is a slim, dual-arm robot with "human-like" flexibility.

(1) Superior dexterity and best-in-class wrist characteristics make the slim, dual-arm robot ideally suited for assembly, part transfer, machine tending, packaging and other handling tasks that formerly could only be done by people.

(2) Highly flexible; 15 axes of motion (7 axes per arm, plus a single axis for base rotation).

(3) Powerful actuator-based design provides "human-like" flexibility and fast acceleration.

(4) Internally routed cables and hoses (6-air, 12-electric) reduce interference and maintenance, and also make programming easier.

(5) Payload is 10 kg per arm; horizontal reach is 720 mm per arm; vertical reach is 1 440 mm per arm; repeatability is ±0.1 mm.

(6) Both robotic arms can work together on one task to double the payload or handle heavy, unwieldy objects. Two arms can perform simultaneous independent operations.

(7) Have the ability to hold part with one arm while performing additional operations with other arm to transfer a part from one arm to the other with no need to set part down.

Vocabulary 词汇

dexterity [dekˈsterəti]	n. 灵巧；敏捷；机敏
cable [ˈkeɪbl]	n. [电]电缆；锚索；连线
optimize [ˈɒptɪmaɪz]	vt. 使最优化，使完善 vi. 优化
acceleration [əkˌseləˈreɪʃən]	n. 加速，促进；[物]加速度
hose [həʊz]	n. 软管
interference [ˌɪntəˈfɪərəns]	n. 干扰，冲突；干涉
maintenance [ˈmeɪntənəns]	n. 维护；保持；生活费用
programming [ˈprəʊɡræmɪŋ]	n. 设计，规划；编制程序

Notes 注释

1. servo actuator　伺服执行器
2. open software architecture　开放的软件架构
3. part transfer　零件转移
4. key benefit　主要优点
5. horizontal reach　水平距离
6. vertical reach　垂直距离
7. independent operation　独立运作
8. additional operation　附加操作；额外操作；附加作业

Part 3　Baxter

For decades, manufacturers have had very few cost-effective options for low volume, high mix production jobs. Then they meet Baxter—the safe, flexible, affordable alternative of fixed automation. Leading companies across North America have integrated Baxter into their workforce, and gained a competitive advantage.

Baxter, as shown in Figure 10-5, is a proven industrial automation solution for a wide range of tasks—from line loading and machine tending, to packaging and material handling. If you walk besides your facility and see monotonous or dangerous tasks, then Baxter is ready to work for your company, doing the monotonous tasks and frees up your skilled human labor exactly.

Figure 10-5　Baxter Robot

The key benefits of Baxter are described as below.

(1) Safe by design. Baxter is safe to operate in production environments, without the need for caging, can save money and valuable floor space.

(2) Easy to integrate. Baxter deploys quickly and connects seamlessly to other automation often without the third party integration.

(3) Can be trained without programming. With Baxter, no traditional programming is required. Instead, it's manually trainable by in-house staff, reducing the time and cost of the third party programmers.

(4) Adaptive. Baxter's compliant arms and force detection let it adapt itself to variable environments, "feeling" anomalies and guiding parts into place.

(5) Flexible and re-deployable. Baxter is flexible for a range of applications and re-trainable across lines and tasks. Baxter can be repurposed quickly across jobs, often

delivering an ROI(return on investment) in under a year.

(6) Affordable and extensible. Baxter's $25 000 base price is feasible for SMEs(small and medium enterprises), and its performance keeps improving through regular software releases.

Some of the applications of Baxter may include, but not limited to, kitting, packaging, loading and unloading, machine tending, material handling, and a few of the accessories are the vacuum cup grippers which can pick up a wide range of objects, especially if they're smooth, non-porous or flat. Electric parallel grippers can pick up rigid and semi-rigid objects of many shapes and sizes. Interchangeable fingers and fingertips maximize flexibility. Mobile pedestal with industrial grade casters makes it easy to move Baxter quickly and safely between workstations.

Baxter is powered by a platform called "Intera", pronounced "in-terra" and named to reflect the robot's interactive production capabilities. Intera provides an easy way to use graphical user interface that in-house staff can master quickly. The platform allows Baxter to be trained by demonstration, using context instead of coordinate to enable non-technical personnel to create and modify programs as needed.

It intelligently handles changing environments, while providing an extensible platform that leverages modern tools such as ROS to maximize relevance and flexibility for the modern workforce.

Vocabulary 词汇

manufacturer [ˌmænjʊˈfæktʃərə]	n. [经]制造商;[经]厂商
workforce [ˈwɜːkfɔːs]	n. 劳动力;工人总数,职工总数
monotonous [məˈnɒtənəs]	adj. 单调的,无抑扬顿挫的;无变化的
cage [keɪdʒ]	vt. 把……关进笼子;把……囚禁起来
integrated [ˈɪntɪɡreɪtɪd]	adj. 综合的,完整的;互相协调的
anomalies [əˈnɒməlɪz]	n. 异常现象;反常现象
re-deployable	可重新配置的
extensible [ɪkˈstensəbl]	adj. 可延长的;可扩张的
accessory [əkˈsesəri]	n. 配件;附件;[法]从犯
interchangeable [ˌɪntəˈtʃeɪndʒəbəl]	adj. 可互换的;可交替的
workstation [ˈwɜːksteɪʃən]	n. 工作站
platform [ˈplætfɔːm]	n. 平台;月台,站台;坛;讲台
interactive [ˌɪntərˈæktɪv]	adj. 交互式的;相互作用的
leverage [ˈlevərɪdʒ]	n. 手段,影响力;杠杆作用

Notes 注释

1. low volume   低容量
2. high mix production   高混合生产
3. industrial automation solution   工业自动化解决方案

4. line loading   线路加载
5. connect seamlessly   无缝连接
6. force detection   力检测
7. vacuum cup gripper   真空夹具
8. electric parallel gripper   电动平行夹具
9. mobile pedestal   移动底座
10. graphical user interface   图形用户界面

Part 4 YouBot

It is small, fun, versatile and has been made for the inventors of the future. The KUKA youBot is a powerful, educational robot that has been especially designed for research and education in mobile manipulation, which counts as a key technology for professional service robotics, as shown in Figure 10-6.

Figure 10-6 YouBot of KUKA

The KUKA youBot consists of two main parts, including an omnidirectional platform and a five degrees of freedom robotic arm with a two-finger gripper.

The KUKA youBot omni-directional mobile platform consists of the robot chassis, four mecanum wheels, motors, power and onboard PCC board, as shown in Figure 10-7. Users can either run programs on this board, or control it from a remote computer. The platform comes with a Live-USB stick with preinstalled Ubuntu Linux and drivers for the hardware.

The KUKA youBot arm has five degrees of freedom(DOF) and a two-finger gripper, as shown in Figure 10-8. If connected to the mobile platform, the arm can be controlled by the onboard PC. Alternatively, the arm can be controlled without the mobile platform by using an own PC connected via Ethernet cable.

Additional sensors can be mounted on the robot.

The KUKA youBot comes with fully open interface and allows the developers to access the system on nearly all levels of hardware control. It further comes with an application programming interface(KUKA youBot API), with interfaces and wrappers for recent robotic frameworks such as ROS or ORCOS, with an open source simulation in Gazebo and with some example codes that demonstrate how to program the KUKA youBot. The platform and the available software shall enable the user to rapidly develop his/her own mobile manipulation applications.

Figure 10-7　The Mobile Platform of YouBot　　　　Figure 10-8　The YouBot Arm

Vocabulary 词汇

versatile ['vɜːsətaɪl]　　　　　　adj. 多才多艺的；万能的；多面手的
chassis ['ʃæsi]　　　　　　　　n. 底盘，底架
wheel [wiːl]　　　　　　　　　n. 车轮；方向盘 vt. 使变换方向
preinstall ['priːɪn'stɔːl]　　　　v. 预设，预安装
driver ['draɪvə]　　　　　　　　n. 驾驶员；驱动程序；起子；传动器
motor ['məʊtə]　　　　　　　　n. 汽车；发动机
Ethernet ['iːθənet]　　　　　　n. [计]以太网
wrapper ['ræpə]　　　　　　　n. 包装材料；[包装]包装纸；书皮
framework ['freɪmwɜːk]　　　n. 框架，骨架；结构，构架
onboard ['ɒnbɔːd]　　　　　　adv. 在船上；在板上 adj. 随车携带的

Notes 注释

1. educational robot　教育机器人
2. professional service robot　专业服务机器人
3. five degrees of freedom robot　五自由度机器人
4. two-finger gripper　双指爪手

Part 5　NAO

NAO (as shown in Figure 10-9) is an autonomous, programmable humanoid robot, developed by Aldebaran Robotics, a French robotics company headquartered in Paris. The robot's development began with the launch of Project NAO in 2004. On 15 August 2007, NAO replaced Sony's robot dog Aibo as the robot used in the RoboCup Standard Platform League (SPL). RoboCup is an international robot soccer competition. NAO was used in RoboCup 2008 and 2009, and the NAOV3R was chosen as the platform for the SPL at RoboCup 2010.

Figure 10-9　NAO Robot

Several versions of the robot have been released since 2008. The NAO academics edition was developed for schools, colleges and universities to teach programming and conduct researches into human-robot interactions. NAO is a groundbreaking teaching aid for use in robotics, system and control, computer science, social science and beyond. Humanoid robots have always fascinated people, especially students. NAO allows them to explore programming, sensor, interaction with people and the environment, and much more. NAO robots have been used for research and education purpose in numerous academic institutions worldwide. By 2015, over 5 000 NAO units are in use in more than 50 countries.

NAO is capable of moving autonomously, having conversation with people, identifying objects and interacting with its environment. Anyone can easily compose programs via the graphical interface of the Choregraphe Software. Students can explore event-based, sequential or parallel programming using the configurable behavior boxes. Users can also create their own behaviors, as well as using Python to write more complex scripts.

NAO can be programmed by either connecting an Ethernet cable between robots and computers, or via a WiFi link. The Webots for NAO Software allows users to simulate their robot programs in a virtual environment that has real world physics. The on-screen NAO can interact with objects in the virtual world, allowing students to test their programs whilst others are using the NAO hardware. The software suite is compatible with Windows, Mac and Linux.

Vocabulary 词汇

version [ˈvɜːʒən]   n. 版本；译本
release [rɪˈliːs]   vt. 释放；发射；让与；允许发表
institution [ˌɪnstɪˈtjuːʃən]   n. 机构；制度
identify [aɪˈdentɪfaɪ]   vt. 确定；鉴定；识别
autonomous [ɔːˈtɒnəməs]   adj. 自治的；自主的；自发的
sequential [sɪˈkwenʃəl]   adj. 连续的；相继的；有顺序的
script [skrɪpt]   n. 脚本；手迹；书写用的字母
interact [ˌɪntərˈækt]   vt. 互相影响 vi. 互相作用

Notes 注释

1. humanoid robot   类人机器人；仿人形机器人
2. NAO academics edition   NAO 学术版

Unit 11　The Outlook for Industrial Robot

Part 1　Current Situation of Industrial Robot

Robot is a piece of automation equipment that combines mechanical, electronic, control, sensing, artificial intelligence and other multi-disciplinary advanced technologies. The current situation of global industrial robot development is described as below.

1. The global market demand of robots continues to grow

The market size of industrial robots and service robots continues to expand. According to the statistics of IFR(International Federation of Robotics), in 2015 the global sales of industrial robots exceeded 240 000 units in the first time, of which Asian sales accounted for about 2/3 of global sales, sales was 144 000 units; European sales was 50 000 units, of which the growth of Eastern European sales reached 29%, it was one of the fastest growing regions in the world; sales in North America reached 34 000 units, representing an increase of 11% over 2014. The total sales of China, Korea, Japan, the United States and Germany, accounted for 3/4 of the global sales. China, the United States, Korea, Japan, Germany, Israel and other countries are more active areas of the industrial robot technology, standards and market development in recent years. In recent years, the sales of industrial robots in the world has grown at a steady rate.

2. Asia-Pacific is the most important market

According to the statistics of IFR, Asia is currently the world's largest region of industrial robot usage amount, accounting for 50% of the world's use of robots, followed by the Americas(including North America, South America) and Europe. 2012-2015, the robot's sales of Asia grew 15% per annum, much higher than the American and African growth rate, which was 6%. In 2015, Asia-Pacific industrial robots were sold more than 140 000 units. In 2014, the new installed capacity of industrial robots in China, Japan, Korea and Thailand accounted for 75% of the total Asian region; respectively, the industrial robot market size of four countries is accounting for 52.4% of global industrial robot sales.

3. The development of industrial robots is highly concentrated

The production and sales of industrial robots are mainly concentrated in Japan, Korea and Germany, the robot holdings and the annual increase of the three countries is in the forefront of the world.

The robot density and ownership of Japan, Korea and Germany is in the world's leading level. According to the statistics of IFR, every 10 000 workers in Japan have 323 industrial robots, 437 units in Korea and 282 units in Germany in 2014. In 2013, the ownership of robots in Japan was 304 000 units, 156 000 in Korea and 168 000 in Germany.

In 2014, the new robot added amount of Japan, Korea, Germany was 30.9% of the global new robots, the market size was 29 000 units, 21 000 units, 20 000 units, respectively. Japanese robot market is mature and its manufacturers are highly competitive internationally. FANUC, NACHI, Kawasaki and other brands continue to lead in the field of microelectronics technology, power electronics technology. Korea's semiconductor, sensors, automated production and other high-end technology have laid the foundation for the rapid development of the robot. German industrial robots in the human-computer interaction, machine vision, machine interconnection and other areas are in the leading level. The German local company KUKA is one of the world's four major manufacturers in the industrial robot field; its annual output is more than 18 000 units.

4. The service robot market is in its infancy

The global service robot market is still in its infancy. The reason for this situation is below. Firstly, the peripheral technology of service robot failed to resolve. Service robot technology is a multi-disciplinary cross integration technology, involving mechanical design, automatic control, bionics, kinematics and other fields, the environment of the diversity, randomness, complexity, the requirements of the task complexity and real-time of its environmental awareness are higher. Secondly, technology of service robot with high value per unit is at low level, and slow development. Such as the requirement of the control movement, fine organization operation and three-dimensional high-definition visual ability of medical robots are high; only a small number of developed countries have the ability to use such technologies.

At present, only some of the national defense robots, household cleaning robots, and agricultural robots can achieve the industrialization in the global service robot market, but higher-tech medical robots and rehabilitation robots are still in the research and development stage. The global personal and home service robot products include home operation robots, recreational robots, disability auxiliary robots and surveillance robots, in which the weeding robots of home operation robots are highly marketed and diversified in product range. For example, Da Vinci Surgical Robots, milking robots and military unmanned aerial vehicles have formed a mature industrial chain.

Vocabulary 词汇

mechanical [mɪˈkænɪkəl]   adj. 机械的；力学的；无意识的
electronic [ɪˌlekˈtrɒnɪk]   adj. 电子的
sense [sens]   n. 感觉；读出；辨向；指向
demand [dɪˈmænd]   vt. 要求；需要；查询 vi. 需要；请求

market ['mɑːkɪt]	n. 市场；行情 vt. 在市场上出售
statistic [stə'tɪstɪk]	n. 统计；统计学；[统计]统计数值
unit ['juːnɪt]	n. [计量]单位
represent [ˌreprɪ'zent]	vt. 代表；表现；描绘；回忆
Israel ['ɪzreɪəl]	n. 以色列；犹太人，以色列人
capacity [kə'pæsəti]	n. 能力；容量；资格，地位；生产力
concentrated ['kɒnsəntreɪtɪd]	adj. 集中的；浓缩的 v. 集中
forefront ['fɔːfrʌnt]	n. 最前线，最前部；活动的中心
density ['densəti]	n. 密度
ownership ['əʊnəʃɪp]	n. 所有权；物主身份
foundation [faʊn'deɪʃən]	n. 基础；地基；基金会；根据；创立
infancy ['ɪnfənsi]	n. 初期；婴儿期；幼年
bionics [baɪ'ɒnɪks]	n. 仿生学
randomness ['rændəmnɪs]	n. 随意；无安排；不可测性
complexity [kəm'pleksɪti]	n. 复杂，复杂性；错综复杂的事物
diversified [daɪ'vɜːsɪfaɪd]	v. 使……多样化
vehicle ['viːɪkl]	n. 车辆；工具；交通工具

Notes 注释

1. artificial intelligence 人工智能
2. current situation 现在的情况
3. steady rate 稳定率
4. installed capacity 装机容量
5. highly concentrated 高度集中
6. microelectronics technology 微电子技术
7. power electronics technology 电力电子技术
8. automated production 自动化生产
9. high-end technology 高端技术
10. human-computer interaction 人机交互
11. machine interconnection 机器互联
12. mechanical design 机械设计
13. automatic control 自动控制
14. control movement 控制运动
15. fine organization operation 良好的组织行为
16. visual ability 视觉能力
17. national defense robot 国防机器人
18. household cleaning robot 家用清洁机器人
19. agricultural robot 农业机器人
20. medical robot 医疗机器人
21. rehabilitation robot 康复机器人

22. home operation robot　家庭操作机器人
23. recreational robot　娱乐机器人
24. disability auxiliary robot　残障辅助机器人
25. surveillance robot　监视机器人
26. weed robot　除草机器人
27. Da Vinci Surgical robot　达芬奇手术机器人
28. milking robot　挤奶机器人
29. military unmanned aerial vehicle　军用无人机

Part 2　Development Trend of Robot

1. Robot and Information Technology In-depth Integration

Big data and cloud storage technology make the robot gradually become the terminal and node of the Internet of things. Firstly, the rapid  development of information technology will be the integration of industrial robots and networks to form a complex production system. A variety of algorithms such as ant colony algorithm, immune algorithm can be gradually applied to the application of robots, so that robots have people's ability to learn. Multiple robots collaboration technology makes a set of production solutions possible. Secondly, the service robot is generally able to achieve remote monitoring through network. Multiple robots can provide more processes, more complex operations of the service; the new mode of operation which human awareness can control robot is also being developed, that is, the use of "thinking" and "willpower" controls the behavior of the robot.

2. Promote Ease of Use and Stability of Robotic Products

With the technologies of standardization of robot structure, integration of joints, self-assembly and self-repair improved, the robot's ease of use and stability has been improved. Firstly, the application of robot has been extended to food, medical, chemical and other broader manufacturing areas, from the more mature cars, electronics industry; service areas and service objects of robots continue to increase; the body of robot has developed to small size, wide application. Secondly, the cost of the robot has rapidly declined. Robot technology and process are becoming more mature. Compared with the traditional special equipment, the initial price gap of robots has been narrow, and in a high degree of personalization, cumbersome process, the robots have higher economic efficiency. Thirdly, man-machine relationship has undergone profound changes. For example, when workers and robots work together, robots can understand human language, graphics, and physical commands through a simple way, eliminate complex worker operations using modular plugs and production components. There is a big safety problem in man-machine collaboration at the existing stage, although light industrial robots with visual and advanced sensors have been developed, there is still a lack of reliable technical specifications for industrial robots.

3. Robots Direct to The Modular, Intelligent and Systematic Way

At present, the global robot products have developed to the modular, intelligent and systematic direction. Firstly, the modular design has improved the problem that the configuration of traditional robots can only be applied to a limited range, industrial robot research has more tended to use reconstruction, modular product design ideas, which help users to solve the contradictions of product variety, specifications, manufacturing cycle with the cost of production. Secondly, in the process of the robot product developing intelligent, the control system of industrial robot has developed to the direction of the open control system integration, the servo drive technology has changed to the direction of the unstructured, multi-mobile robot system, robot collaboration has been not only the coordination of control, but been the coordination of the organization and control of the robot system. Thirdly, the industrial robot technology continues to extend, the current robot products are being embedded to construction machinery, food machinery, laboratory equipment, medical equipment and other traditional equipment.

4. The Market Demand for New Intelligent Robots Increases

The demand for new intelligent robots, especially those with intelligence, flexibility, cooperation and adaptability, continues to grow. Firstly, the fine work ability of the next generation of intelligent robots is further enhanced; the ability to get adapted to the outside world has been increasing. In terms of the fine operating capability of the robot, the Boston Consulting Group survey showed that robots entering factories and laboratories recently had distinct characteristics that could accomplish fine work such as assembling tiny parts; pre-programmed robots don't need expert monitoring. Secondly, the market demand for robot flexibility continues to increase. Recently, Renault used a batch of screwing robots of which the load is 29 kg, the only 1.3 m long arm in which there are six rotary joints embedded can be flexible operation. Thirdly, the demand of interpersonal skills continues to grow. In future, robots are able to perform tasks near workers. A new generation of intelligent robots use SONAR, cameras, or other technologies to sense if there are workers in the work environment, and if there is a collision possible, they may slow down or stop functioning.

Vocabulary 词汇

terminal [ˈtɜːmɪnəl]   n. 末端；终点；终端机 adj. 末端的
willpower [ˈwɪlpaʊə]   n. 意志力；毅力
standardization [ˌstændədaɪˈzeɪʃən]   n. 标准化；[数]规格化；校准
self-repair [ˈselfrɪˈpeə]   n. 自行修复
personalization [ˈpɜːsənəlaɪzeɪʃn]   n. 个性化
modular [ˈmɒdjʊlə]   adj. 模块化的；模数的
systematic [ˌsɪstɪˈmætɪk]   adj. 系统的；体系的；有系统的
reconstruction [ˌriːkənˈstrʌkʃən]   n. 再建，重建；改造；复兴
screw [skruː]   vt. 旋；拧；压榨 n. 螺旋；螺丝钉
intelligence [ɪnˈtelɪdʒəns]   n. 智力；情报工作；情报机关
cooperation [kəʊˌɒpəˈreɪʃən]   n. 合作；协作；协力
SONAR [ˈsəʊnɑː]   n. 声波水下探测系统；声呐
camera [ˈkæmərə]   n. 摄像机；照相机
collision [kəˈlɪʒən]   n. 碰撞；冲突；抵触

Notes 注释

1. big data　大数据
2. cloud storage technology　云存储技术
3. ant colony algorithm　蚁群算法
4. immune algorithm　免疫算法
5. remote monitoring　远程监控
6. integration of joints　集成一体化关节
7. initial price gap　初始价格差距
8. cumbersome process　烦琐的过程
9. modular plug　模块化插头
10. technical specification　技术规格
11. systematic way　系统的方式
12. laboratory equipment　实验室设备
13. new intelligent robot　新智能机器人

Part 3 Social Issues Caused by Application of Robot

At present, most robots are mainly used in the industrial field, but for the demand of people's daily life, a variety of robots have gradually come to us. In the context of the wider use of robots, robots may profoundly change our life, and the resulting ethical issues are increasingly pressing in front of us. For more than ten years, robotics ethical research is also a hot topic in academia.

1. Military Robots: to Cherish Life or Ruthless Killing

The use of robots in the military is one of the important driving forces in the development of robot technology. At present, the world is actively researching military robots including ground robots, air robots, underwater robots and space robots. Relative to human soldiers, military robots can fight in more harsh environments, and absolutely obey orders. They do not need to be repeatedly trained; the advantages are obvious. Military robots charge forward instead of human soldiers in the battle fields, and they can greatly reduce the casualties of personnel. Therefore, the use of military robots seems to be an important means of cherishing human life.

However, military robots under modern scientific and technical armed forces can have more destructive power than human soldiers. What's more, robots may have no sympathy for humans, and their strong lethality can make them true cold-blooded "killer" machines. Of course, people can solve this problem by agreeing that "robots and robots are fighting, and robots do not fight with humans," but this solution is difficult to achieve. Therefore, the ethical problem of military robots is one of the most discussed contents in robotics research.

2. Child Care Robots: to Reduce the Burden or Shirk Responsibility

Child care robots are widely valued in Korea, Japan and a few European countries. Child care robot has many functions including video games, voice recognition, facial recognition and communication. They are equipped with visual and auditory monitor; they can move, and deal with some problems on their own. When the child leaves the specified range, child care robots will alarm.

Are child care robots liked by children? Experiments show that child care robots can be used as children's intimate playmates and they are more important especially for young children who lack little partners, because they are much better than the average toys, and can make children happier. More and more child care robots have entered into the ordinary families.

There is no doubt that child care robots can reduce the burden on parents, so that they can have more free time. However, for the healthy growth of children, the care of parents can't be replaced, and the robot can only play a supporting role. If the children are taken to the robots for care more often, then it may affect the normal development of their psychology and emotion.

3. Robot to Accompany: Getting Rid of Loneliness or Keeping Away from Community

Population aging is the universal law of population development in many countries of the world, our country is no exception. From a single family's perspective, the help-old robots can reduce the time of family care for the elderly, thereby reducing the burden on the family. Robots can take place of human care workers, which can reduce the workload of human care workers, reduce the social demand for human care workers, thereby reducing the social burden. Moreover, compared with human care workers, help-old robots have some unique advantages. For example, robots can serve the elderly 24 hours a day. Another example, a wide range of robots can meet the different needs of the elderly, significantly improve the quality of life of the elderly, and even can be customized according to the needs of the elderly, but human care workers are difficult to do this. In addition, the help-old robots can also interact, entertain with the elderly, and even meet the emotional needs of the elderly in a certain extent to reduce the loneliness of the elderly living alone. Everyone nearly has a positive and optimistic attitude towards the use of help-old robots.

However, the help-old robots may also lead to a series of ethical issues, such as the elderly's privacy, freedom, dignity and other issues. One of the most prominent problems may be the contact between elderly and family, elderly and society. If much of the work which should be done by human is given to the robots to do, it is possible to greatly reduce the exchanges and contacts between elderly and society. Studies have shown that more social exchanges and interactions can extend the longevity of older people who need long-term care. There is no doubt that the elderly also have a strong emotional need, whether this need is met, and how much extent it is to be met, for the health of the elderly it is essential.

4. Robot Technology and Its Application: Free development or Ethical Regulation

Some scientists believe that the nature of science and technology is neutral; science and technology research is without a restricted area and the consequences of technology applications have nothing to do with technology itself. How to use it, whether it is to bring happiness or to bring disaster, all depends on humans, but doesn't depend on tools. Some scholars believe that the existing ethical considerations on robots are mainly based on the theoretical research; what the reality will be is inconclusive, we should firstly develop related science and technology, there is no need to slow down the pace because of some possible negative effects. If the research of humanities and social sciences is based on the full development of existing technologies and their social effects, it will produce "post-cultural phenomenon" and "system vacuum", which is obviously unfavorable to the development of human civilization. Therefore, the research of humanities and social sciences must have a certain predictability and forward-looking to take precautions.

In view of the profound social impacts that robotics may have, having ethical regulation is as equally important as paying attention to technological development. The ethical regulation of robotic technology is to standardize, institutionalize and materialize the principles of robot ethics, and its purpose is to protect the interests of the whole and the individual.

If there is a far-reaching impact of the "robot revolution" in the twenty-first century, then this revolution will not only be science and technology revolution, but also a social and ethical revolution. The widespread use of robots is unavoidable, and we don't need to focus on the questions that whether the robots should be used, but how to use them better. In order to make robots better for social services, the development of technology is of course necessary, while ethical considerations are also crucial.

Vocabulary 词汇

context ['kɒntekst]	n. 环境;上下文;来龙去脉
research [rɪ'sɜːtʃ]	n. 研究 vi. 研究 vt. 研究
academia [ˌækə'diːmiə]	n. 学术界;学术生涯
harsh [hɑːʃ]	adj. 严厉的;严酷的;刺耳的
session ['seʃən]	n. 会议;(法庭的)开庭
auditory ['ɔːdɪtəri]	n. 听众;礼堂 adj. 听觉的;耳朵的
monitor ['mɒnɪtə]	n. 监视器;监听器;监控器
psychological [ˌsaɪkə'lɒdʒɪkəl]	adj. 心理学的;精神上的
emotional [ɪ'məʊʃənəl]	adj. 情绪的;易激动的;感动人的
entertainment [ˌentə'teɪnmənt]	n. 娱乐;消遣;款待
predictability [prɪˌdɪktə'bɪlɪti]	n. 可预测性;可预言性
forward-looking ['fɔːwədˌlʊkɪŋ]	adj. 有远见的;向前看的
standardize ['stændədaɪz]	vt. 使标准化;用标准检验
institutionalize [ˌɪnstɪ'tjuːʃənəlaɪz]	vt. 使……制度化
materialize [mə'tɪərɪəlaɪz]	vt. 使具体化;使有形 vi. 实现;成形;突然出现
ethical ['eθɪkəl]	adj. 伦理的;道德的

Notes 注释

1. industrial field   工业领域
2. ethical issue   伦理道德问题
3. military robot   军事机器人
4. ground robot   地面机器人;陆地机器人
5. air robot   空中机器人
6. underwater robot   水下机器人
7. space robot   空间机器人
8. child care robot   儿童看护机器人
9. video game   视频游戏
10. voice recognition   语音识别
11. facial recognition   面部识别
12. help-old robot   助老机器人

Part 4　Latest Industry Data of Industrial Robot

At present, the scale of the global robot market continues to expand, and the industrial & special robot market is growing steadily. The application of technology innovation around the biomimetic structure, artificial intelligence and human-machine collaboration continues to deepen. The application of products in the fields of education escort, medical rehabilitation, dangerous environment and other fields continues to expand. The enterprise prospective layout and investment merger and acquisition are extremely active, and the global robot industry is coming to a new round of growth.

Global sales of industrial robots reached the new record of 387 000 units in 2017. That is an increase of 31 percent compared to the previous year (2016: 294 300 units). China saw the largest growth in demand for industrial robots, up to 58 percent. Sales in the USA increased by 6 percent and in Germany by 8 percent compared to the previous year. These are the initial findings of the World Robotics Report 2018, published by the IFR. The estimated worldwide annual shipments of industrial robots 2006—2017 is shown in Figure 11-1.

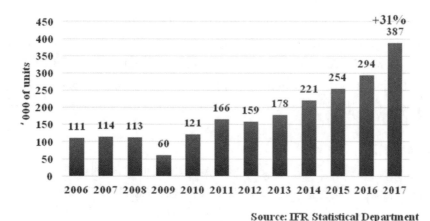

Source: IFR Statistical Department

Figure 11-1　Record Growth of Industrial Robots

Broken down by industry, the automotive industry continues to lead global demand for industrial robots. In 2017, around 125 500 units were sold in this segment—equivalent to growth of 21 percent. The strongest growth sectors in 2017 were the metal industry (+55 percent), the electrical/electronics industry (+33 percent) and the food industry (+19 percent).

In terms of sales volume, Asia has the strongest individual markets: China installed around 138 000 industrial robots in 2017, followed by Japan with around 46 000 units and Korea with around 40 000 units. In the Americas, the USA is the largest single market with around 33 000 industrial robots sold, and in Europe it is Germany with around 22 000 units sold. The estimated worldwide annual supply of industrial robots at year-end main markets 2015—2017 is shown in Figure 11-2.

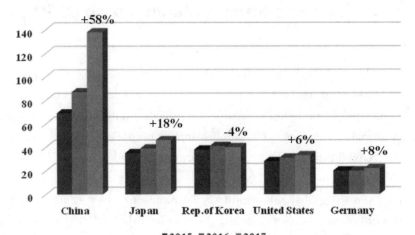

Figure 11-2   Annual Supply of Industrial Robots

"The growth of industrial robots continues at an impressive pace worldwide." says JunjiTsuda, President of the International Federation of Robotics. "Key trends such as digitalisation, simplification and human-robot collaboration will certainly shape the future and drive forward rapid development."

In the course of digitalisation, real production is becoming increasingly connected with the virtual data world, opening up completely new possibilities for analysis—right through to machine learning. Robots will acquire new skills through learning processes. At the same time, the industry is working to simplify the handling of robots. In the future industrial robots should be easier and faster to program using intuitive procedures.

Vocabulary 词汇

Munich ['mjuːnɪk]	n. 慕尼黑(德国城市,巴伐利亚州首府)
innovation [ˌɪnəˈveʃən]	n. 改革,创新;新观念;新发明;新设施
biomimetic [ˌbaɪɒmɪˈmetɪk]	n. 仿生的;拟生态的
artificial [ˌɑːtɪˈfɪʃl]	adj. 人工的;人造的;人为的;虚假的,非原产地的
intelligence [ɪnˈtelɪdʒəns]	n. 智力;聪颖
previous [ˈpriːviəs]	adj. 以前的;先前的;(时间上)稍前的
initial [ɪˈnɪʃl]	adj. 最初的;开始的;首字母的
international [ˌɪntəˈnæʃnəl]	adj. 国际的;两国(或以上)国家的
federation [ˌfedəˈreɪʃn]	n. 联盟,联合会;联邦,联邦政府
escort [ˈeskɔːrt]	n. 护送者;陪护;vt. 护送,护卫
segment [ˈsegmənt]	n. 环节;部分,段落
equivalent [ɪˈkwɪvələnt]	adj. 相等的,等效的;等价的,等积的
individual [ˌɪndɪˈvɪdʒuəl]	adj. 个人的;个别的;独特的

digitalisation [dɪdʒɪteɪlaɪˈzeɪʃən] *n.* 取样频率；数字化
simplification [ˌsɪmplɪfɪˈkeɪʃn] *n.* 单纯化，简单化
collaboration [kəˌlæbəˈreɪʃn] *n.* 合作，协作；通敌，勾结

Notes 注释

1. human-machine collaboration　人机协作
2. the automotive industry　汽车行业
3. the metal industry　金属行业
4. International Federation of Robotics　国际机器人联合会
5. the previous year　上一年
6. In terms of…　根据……；就……而言

Unit 12   Intelligent Manufacturing and Global Robot Development Program

Part 1   Industry 4.0

The fourth industrial revolution, also known as Industry 4.0, the factory of the future, the smart factory, the industrial Internet, is quickly emerging and affecting our lives in new ways. It is helpful to streamline robots and their automated operations, while also optimizing costs, turning the potential for enterprise-wide automated business transformation into a reality.

The fourth industrial revolution is the new fusion of automation and data exchange that has created the perfect environment for these "smart factories". It is a new connectivity that combines and networks all of the advancements in our technology world: intelligent systems(cyber-physical systems), the Internet of Things, and cloud computing. All of these items are programmed to work together to enable humans and machines to cooperate and communicate, with each other or among themselves, in real time and even remotely. All of the devices, systems, and people in the production chain are linked and able to deliver data in the right form whenever and wherever necessary.

Industry 4.0 is continuing to prove that humans are unstoppable in their innovative solutions towards a more effective and intelligent society. Since the beginning, humans have been on a constant search to continue innovating and finding new solutions that make creating products more affordable, smarter, and also make our lives a little easier. The revolutions of the past(water and steam power, electric power, and digital power) have truly transformed the society, governing structures, and human identities in which they are found. All of these individual revolutions have aided manufacturing businesses to continue their unvarying searches for better and more economically feasible solutions.

The first industrial revolution began to use fossil fuels and mechanical power for energy sources. The second industrial revolution brought electrical distribution, wired and wireless, and new forms of power generation. The third revolution brought rapid advances in computing power, enabling new ways of generating, processing, and sharing informations. It also used electronics and IT systems to achieve further automation of manufacturing. All of these revolutions were driven not only by researchers, inventors, and designers, but also more importantly by the people who adopted and employed them in their everyday lives. The arrival of Industry 4.0 really has been a reflection of the desires of its citizens and their choices.

The diagram of four industrial revolution is shown in Figure 12-1.

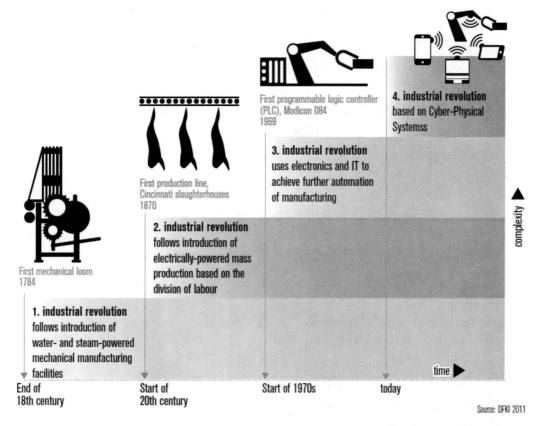

Figure 12-1　The Diagram of Four Industrial Revolution

So the new industrial revolution we see today is really of no surprise. Reliant on all revolutions that existed before it, the fourth industrial revolution is offering entirely new capabilities for people and machines and ways in which technology becomes embedded within societies. The level of manufacturing agility can make it possible to connect customers' needs with the ability of the company to deliver a product, potentially on demand. Manufacturers will be able to listen and then act to better adapt to the consumers' demands as they can access real-time intelligence and analytics, on-demand.

If we have the courage to take collective responsibility for the changes underway, and the ability to work together to raise awareness and shape new narratives, we can embark on restructuring our economic, social and political systems to take full advantages of emerging technologies.

It is clear that the fourth industrial revolution is here to stay and if you are in the manufacturing business, sitting on the sidelines is no longer a good idea. Robotic technology is well on its way to revolutionize the way to the success of a business; the benefits are increasingly clear and attainable to those of all sizes.

Vocabulary 词汇

emerging [ɪˈmɜːdʒɪŋ]　　　　　　v. 形成；浮现；显露
fusion [ˈfjuːʒən]　　　　　　　n. 融合；熔化；熔接；融合物；[物]核聚变
remotely [rɪˈməʊtli]　　　　　　adv. 遥远地；偏僻地；(程度)极微地，极轻地
innovative [ˈɪnəˌveɪtɪv]　　　　adj. 革新的，创新的；新颖的
distribution [ˌdɪstrɪˈbjuːʃən] n. 分布；分配
wired [ˈwaɪəd]　　　　　　　　adj. 接有电线的；以铁丝围起的 v. 以金属丝装；打电报给
generating [ˈdʒenəreɪtɪŋ]　　　adj. 产生的 n. 产生 v. [计]生成
inventor [ɪnˈventə]　　　　　　n. 发明家；[专利]发明者
designer [dɪˈzaɪnə]　　　　　　n. 设计师；谋划者
reflection [rɪˈflekʃən]　　　　　n. 反射；沉思；映象
citizen [ˈsɪtɪzən]　　　　　　　n. 公民；市民；老百姓
capability [ˌkeɪpəˈbɪləti]　　　n. 能力；功能；性能
agility [əˈdʒɪləti]　　　　　　　n. 敏捷；灵活；机敏
shape [ʃeɪp]　　　　　　　　　n. 形状；模型 vt. 形成；使成形 vi. 形成；成形；成长

Notes 注释

1. industrial revolution　　工业革命
2. factory of the future　　未来工厂
3. smart factory　　智慧工厂
4. industrial Internet　　工业互联网
5. intelligent system　　智能系统
6. Internet of Things　　物联网
7. cloud computing　　云计算
8. real time　　即时的
9. production chain　　生产链
10. steam power　　蒸汽动力
11. electric power　　电力
12. digital power　　数字动力
13. governing structure　　管理结构
14. manufacturing business　　制造业务
15. fossil fuel　　化石燃料
16. mechanical power　　机械动力
17. energy source　　能源
18. power generation　　发电
19. computing power　　计算能力
20. share information　　共享信息
21. take collective responsibility for　　承担集体责任
22. raise awareness　　提高意识
23. emerging technology　　新兴技术

Part 2 Core Technology of Intelligent Manufacturing

Intelligent manufacturing (IM) is intelligent man-machine integration system worked out by smart machines and human experts, It can do intelligent activities in the manufacturing process, such as analysis, inference, judgement, conception, decision, etc, through people, between people and machines, machine and the synergy, to expand, extend or partially replace human experts in the manufacturing process of mental work.

Intelligent manufacturing is the inevitable outcome of the mechanization, automation and information technology which are applied to the mature stage.

Intelligent manufacturing includes the following core technologies.

1. Cyber-physical Systems(CPS)

It is basically a catch-all term for talking about the integration of smart, Internet-connected machines and human labor. Factory managers are not simply reforming the assembly line, but actively creating a network of machines that not only can produce more with fewer errors, but can autonomously alter their production patterns in accordance with external inputs while still retaining a high degree of efficiency.

2. Artificial Intelligence(AI)

It is a theory, method, technique and application system which is used to research and develop how to simulate, extend and expand human intelligence. It seeks to understand the essence of intelligence and produce a new intelligent machine that can react in a way that is similar to human intelligence. The research in this field includes robotics, language recognition, image recognition, natural language processing and expert systems.

3. Augmented Reality Technology(AR)

It is using technology to superimpose information on the world we see. For example, images and sounds are superimposed on what the user sees and hears. Picture the "Minority Report" or "Iron Man" style of interactivity. This is rather different from virtual reality. Virtual reality means computer-generated environments for you to interact with, and being immersed in. Augmented reality adds to the reality you would ordinarily see rather than replacing it. Augmented reality is often presented as a kind of futuristic technology, but it's been around in some form for years, if your definition is loose. For example, the heads-up displays in many fighter aircrafts as far back as the 1990s would show information about the attitude, direction and speed of the plane, and only a few years later they could show which objects in the field of view were targets.

4. The Internet of Things(IoT)

The Internet of things is the Internet which everything is connected, it real-timely monitors, connects, interacts with objects or all kinds of information needed, through a variety of information sensing devices, it forms a huge network combined with the Internet. Its purpose is to realize the content between objects, people, all of the items and the network, easy to identify, manage, and control.

5. Industrial Big Data(IBD)

IBD has been applied in industrial field, in order to make the equipment data, activity data, environmental data, service data, business data, market data and up-down stream industry chain data which are originally isolated, mass, diversity mutually connect, then realize the connection of human and human, material and material, and between the people and things, especially to realize the connection of end users with the process of manufacturing, service. Through the new processing mode, according to the real-time requirement of the business scene, it realizes the mutual transformation of data, information and knowledge, it has a stronger decision-making, insight found and process optimization ability. Compared with other big data, industrial big data has a greater characteristic of profession, relevance, process, timing and analyticity, etc.

Vocabulary 词汇

synergy [ˈsɪnədʒi]　　　　　　　n. 协同;协同作用;增效
simulate [ˈsɪmjʊleɪt]　　　　　vt. 模仿;假装;冒充 adj. 模仿的;假装的
essence [ˈesens]　　　　　　　n. 本质,实质;精华;香精
superimpose [ˌsuːpərɪmˈpəʊz]　vt. 添加;重叠;附加;安装
diversity [daɪˈvɜːsəti]　　　　　n. 多样性;差异
optimization [ˌɒptɪmaɪˈzeɪʃən]　n. 最佳化,最优化
profession [prəˈfeʃən]　　　　　n. 职业;专业;声明;宣布
relevance [ˈrelɪvəns]　　　　　　n. 关联;适当;中肯
process [ˈprəʊses]　　　　　　　vt. 处理;加工 n. 过程;进行 vi. 列队前进 adj. 经过特殊加工
timing [ˈtaɪmɪŋ]　　　　　　　　n. 定时;调速 v. 为……安排时间
analyticity [ˌænəlɪˈtɪsəti]　　　　n. 分析性;[数]解析性

Notes 注释

1. intelligent manufacturing　智能制造
2. man-machine integration system　人机一体化系统
3. cyber-physical system　网络物理系统
4. artificial intelligence　人工智能
5. language recognition　语言识别
6. image recognition　图像识别
7. natural language processing　自然语言处理
8. expert system　专家系统
9. augmented reality technology　增强现实技术
10. virtual reality　虚拟现实

11. the Internet of things   物联网
12. industrial big data   工业大数据
13. up-down stream industry chain   上下游产业链
14. end user   最终用户
15. processing mode   处理模式

Part 3 Wisdom Factory

In recent years, China has constantly introduced targeted measures to promote the development of intelligent manufacturing, and pointed out the new direction of development for the traditional manufacturers—wisdom factory. Wisdom factory will build the communication bridge between a product and the manufacture, for undertaking the implementation of intelligent manufacturing. Therefore, the future of the intelligent manufacturing will be the wisdom factory.

The so-called wisdom factory set a variety of emerging technologies and intelligent systems in one of human chemical plants based on the digital factory. Wisdom factory has improved the control of the production process, reduced human intervention of the production line, timely and accurately collected the operational data, therefore it can enhance the core competitiveness, improve production efficiency and reasonably arrange production and so on.

From the definition we can see, the realization of intelligent factories can't be separated from the support and application of emerging technologies. For example, without the extensive use of advanced sensors, the wisdom factory is difficult to be called smart. The development of advanced sensors relies on the progress of microprocessors and artificial intelligence technology.

The application of software engineering assistance system is also one of the basic elements of intelligent factory. Software engineering assistance system is a based on knowledge, highly integrated intelligent software system, with digital expression, information access, processing knowledge and other capabilities. The wisdom factory will gradually replace the traditional single mode of work.

Of course, to realize the wisdom factory, we need to focus on breaking through machine tools, processes, production control and other key technologies.

Take machine tools as an example, CNC machine tools are the most basic part of the wisdom factory. Therefore, the construction of the wisdom factory can't be separated from the advance of the wisdom machine tools. Intelligent machine tools can independently collect data and determine the operation status, automatically detect, induce, simulate the intelligent decision-making of the goals, so that the machine operation is in the best condition.

In addition, the development of intelligent logistics is also a key factor of the fact that intelligent factories rapidly spread. For intelligent factories, intelligent logistics is a "recycling system", which needs continuously to transport production-related resources. Fortunately, China vigorously developed intelligent logistics in recent years; information and intelligent construction have made great progress.

In general, the wisdom factory is the ultimate goal of the modern plant being information technology development, and also an important step to achieve intelligent manufacturing.

Vocabulary 词汇

digital ['dɪdʒɪtl]	adj. 数字的；手指的 n. 数字
definition [ˌdefɪ'nɪʃən]	n. 定义；[物]清晰度；解说
application [ˌæplɪ'keɪʃən]	n. 应用；申请；应用程序；敷用
extensive [ɪk'stensɪv]	adj. 广泛的；大量的；广阔的
advanced [əd'vænst]	adj. 先进的；高级的；晚期的；年老的 v. 前进；增加；上涨
elements ['elɪmənts]	n. 基础；原理
access ['ækses]	vt. 使用；存取；接近 n. 进入；使用权；通路
CNC	abbr. 电脑数字控制
status ['steɪtəs]	n. 地位；状态；情形；重要身份
microprocessor [ˌmaɪkrəʊ'prəʊsesə]	n. [计]微处理器

Notes 注释

1. targeted initiative 有针对性的措施
2. wisdom factory 智慧工厂
3. human chemical plant 人性化工厂
4. production line 生产线
5. core competitiveness 核心竞争力
6. artificial intelligence 人工智能
7. software engineering 软件工程
8. machine tool 机床
9. production control 产品控制
10. recycling system 回收系统

附录1 参考译文

第1单元 机器人概述

第1部分 关于机器人

机器人在工作和家庭中被广泛应用。这些机器可以完成对人而言太过危险的工作,帮助做家务,或者只是用来娱乐。

什么是机器人和机器人学?

"机器人"一词是捷克作家卡雷尔·查培克于1920年在其剧本《罗萨姆的万能机器人》中最早向公众提出的,意指"被动工作与劳动"。戏剧开始于一家使用机器人的工厂。这些机器人被描述为高效的,但没有感情、无法思考、无法自卫,如图1-1所示,这是最早的工业机器人设想。

如今,机器人指的是所有能通过自动或遥控替代人类执行工作或从事其他活动的人造机械。机器人是可以用来从事工作的机械,部分机器人可以自主作业,其余的机器人必须有人操控。

机器人学是一门研究、设计和使用机器人系统的技术学分支。机器人学与电子学、工程学、机械学和软件科学息息相关,这些技术使机器人能够在危险的环境或制造过程中取代人类工作,或者在外观、行为和认知方面模仿人类。如今,许多机器人的灵感都来自大自然,这有助于仿生机器人技术的研究。

机器人定律

机器人定律是一整套的法律、规则或原则,旨在为设计具有一定自主程度的机器人提供基本的框架。最著名的是1942年美国作家艾萨克·阿西莫夫的"机器人三原则"(通常简称为"三原则"或"阿西莫夫定律")。这三条定律如下所述。

第一定律:机器人不得伤害人类个体,或者目睹人类个体将遭受危险而袖手旁观;

第二定律:机器人必须服从人给予它的命令,当该命令与第一定律冲突时例外;

第三定律:机器人在不违反第一、第二定律的情况下要尽可能保护自己的生存。

在后来的小说中,阿西莫夫补充了第零定律,来扩展其他定律。

第零定律:机器人不得伤害人类的整体利益,或通过不采取行动,让人类利益受到伤害。

这四条定律被广泛用于定义现实和科幻中的机器人准则。

第 2 部分　机器人的种类及应用

随着科技的不断进步,机器人学成为一个快速发展的领域。如今,机器人可作为玩具、吸尘器和可编程宠物出现在家庭中。机器人也已经成为工业、医学、科学、空间探索、建筑、食品包装甚至手术等许多方面的重要组成部分。

机器人有许多种类,例如:移动机器人、工业机器人(操纵)、服务用机器人、教育机器人、模块化机器人、协作机器人等,如图 1-2 和图 1-3 所示。它们被用于不同的环境和用途。机器人已经取代人类执行重复和危险的工作,这些工作都是人类不愿意做的或者由于尺寸限制而做不到的,或者是需要在外太空或海底等极端环境中完成的。

总而言之,机器人可用于以下两类工作。

(1) 机器人做得比人类好的工作。这里,使用机器人可以提高生产率、精确度和耐久性。

(2) 人类做得比机器人好,但出于某种原因需要机器人代替人类做。这里,机器人将我们从肮脏、危险和枯燥的工作中解放出来。

第 3 部分　关于工业机器人

定义和应用

工业机器人是用于生产制造的机器人系统。ISO8373 中,工业机器人被定义为"一个自动控制的、可重复编程的、多用途的、有三个或更多个轴的可编程控制器,可以在工业自动化应用中固定或移动使用"。

美国机器人协会(RIA)将机器人定义为:"一种可用于移动材料、零件、工具或专用装置的,通过可编程动作来执行种种任务的,具有编程能力的多功能机械手(manipulator)。"

工业机器人有助于物料搬运,可提供接口。机器人的典型应用包括焊接、喷涂、组装、PCB 板取放、包装、贴标、码垛、检测,机器人都以很高的耐久性、高速性和精确度实现这些工作。

工业机器人的历史

1959 年,乔治·德沃尔(George Devol)发明了第一台工业机器人 Unimate,如图 1-4 所示。1961 年,该机器人被用于通用汽车公司的生产线。在美国新泽西州的一家工厂中,该机器人被用来从压铸机中提取红热的门把手和其他的汽车零件,并将其堆放起来。因此,第一台 Unimate 是材料处理机器人。它最鲜明的特点是夹具,夹具的使用消除了工人接触由熔融金属制成的汽车零件的工作需要。

该机器人很快被效仿,用于焊接和其他应用,它们承担了从装配线运输压铸件的工作,并将这些部件焊接在汽车车身上。这项工作对工人而言很危险,一不小心他们可能会烟雾中毒或失去肢体。

商业的工业机器人

1962 年,AMF 公司生产了 Verstran 机器人,同 Unimate 一样,该机器人也是商用的工业机器人,并出口到世界各国,在全球范围内掀起了机器人热潮。

20 世纪 70 年代,许多公司开始了机器人业务,并制造了他们的第一台工业机器人。如 Nachi,KUKA,FANUC,安川,ASEA(ABB 的前身),OTC。

1978 年,Unimation 在通用汽车公司的支持下推出了 PUMA(用于装配的可编程通用机器)机器人,如图 1-5 所示。如今,PUMA 仍然工作在生产第一线,许多工业机器人的研究都基于该机器人的模型和对象。

1979 年,OTC 原本是一家焊接设备供应商,后来扩大成为 GMAW 日本汽车市场的供应商,之后 OTC(日本)推出了第一代专用的电弧焊机器人。

1980 年,工业机器人行业开始快速增长,每个月都会有新的机器人或公司面世。

第 4 部分　工业机器人的组成及应用

最常用的机器人有关节型机器人、SCARA 机器人、Delta 机器人和笛卡尔坐标机器人(龙门式机器人或 X-Y-Z 机器人)。

组成部分

典型的工业机器人由机械臂、控制系统、示教器、末端执行器以及一些其他外围设备组成,如图 1-6 所示。

机器人的机械臂主要是移动工具的部分,不是每个工业机器人的机械臂都类似于手臂,不同类型的机器人有不同的结构。控制系统类似于机器人的大脑;示教器构成了用户环境,通常仅在编程时使用;末端执行器是为特定任务(例如焊接或喷涂)而设计的装置。

种类

下面介绍工业机器人的主要类型。

1. 笛卡尔坐标机器人(直角坐标机器人)/龙门式机器人

用于拾取和放置工作、点胶、组装操作和焊接。该机器人的手臂有三个柱形关节,轴线与笛卡尔坐标一致。

2. 圆柱形机器人

用于组装操作、操作机床、点焊和操作压铸机。该机器人的轴线形成圆柱坐标系。

3. 球形机器人

用于操作机床、点焊、压铸、保养机器。该机器人的轴线形成极坐标系。

4. SCARA 机器人

用于拾取和放置工作、点胶、装配和操作机床。该机器人具有两个平行的旋转关节。

5. 铰接式机器人(关节机器人)

用于装配作业、压铸、焊接和喷涂。该机器人的手臂至少有三个旋转关节。

6. 并联机器人

它是一种手臂具有并列的棱柱或旋转关节的机器人。

应用

这里介绍工业机器人的四个主要应用。

1. 机器人搬运

如图 1-7 所示,物料搬运是工业机器人在全世界最流行的应用,包括机器人机器管理、金属加工和塑料成型的各种操作。在过去的几年中,随着协作机器人的引入,该部分市场持续增大。

2. 机器人焊接

如图 1-8 所示,这部分主要包括点焊和电弧焊接,主要用于汽车工业。越来越多的小工厂开始将焊接机器人引入生产。事实上,随着机器人价格的下降和市面上现有的各种工具的增多,实现自动化焊接工艺越来越容易。

3. 机器人装配

如图1-9所示,装配操作包括固定、压接、插入、拆卸等。在过去几年,尽管其他机器人的应用增加,这类机器人的应用似乎有所减少。机器人应用多样化是因为引入了不同的技术,例如力矩传感器和触觉传感器,它们给机器人带来更多的感觉。

4. 机器人点胶

如图1-10所示,我们在这里讨论的是喷漆、上胶、涂胶、喷涂等。只有4%的机器人用于点胶,机器人的流畅性造就了可重复的、精确的点胶工艺。

第2单元 工业机器人概述

第1部分 工业机器人的主要部分

机器人可以由各种材料制成,包括金属和塑料。工业机器人由操作机、控制器、示教器、机器手部分四部分组成,如图2-1所示。

(1)操作机,即机械臂,是一种机械手臂,通常是可编程的,具有与人类手臂相似的功能。机械臂可能是机器人机构的总和,也可能是更复杂的机器人的一部分。机械臂通过接头连接,允许旋转运动或平移(线性)位移。机械臂的末端是末端执行器。

(2)控制器也称为由计算机程序运行的"大脑"。通常,程序非常详细,因为它为机器人的运动部件提供命令。

(3)示教器也称为示教盒,是与控制器相连的人机交互界面,可由操作员手持操作。

(4)机器手,即末端执行器,或手臂末端工具(EOT),可根据需要进行任意设计来完成各类作业,如焊接、夹持、旋压等,具体取决于应用。末端执行器通常非常复杂,需要与要处理的产品相匹配,并且通常能够同时拾取一系列产品。他们可以利用各种传感器,协助机器人系统进行定位、搬运和放置产品。

所有这些部件联合工作,来控制机器人的操作。

第2部分 基本术语

1. 刚体

在物理学上,理想的刚体是形变情况可以被忽略的固体。换言之,无论何时施加外力,刚体的任意两个给定点之间的距离都保持不变。

2. 旋转关节

旋转关节(也称为销接头或铰接头)是用于机构中的具有一个自由度的运动副。旋转关节提供单轴旋转功能,可以用于门铰链、折叠机构,以及其他单轴旋转的装置等许多场合。

3. 运动副

运动副是两个物体之间的连接,它们对相对运动施加约束。

4. 关节机器人

关节机器人是一种使用旋转关节来接近其工作空间的机器人。通常,接头被布置成"链",使得一个关节在链中进一步支撑其他关节。

5. 连续路径

一种通过输入或命令指定每个点沿着期望的路径运动的控制方案,该路径由机器手关节的协调运动来控制。

6. 运动学

机器人中刚性构件和关节的实际排列决定了机器人可能的运动。机器人运动学的种类包括铰接、笛卡儿、并联和 SCARA。

7. 运动控制

对于一些应用,例如简单的取放、组装,机器人仅需要重复地返回到有限的预示教点位。对于更复杂的应用,例如焊接和精加工(喷漆),运动必须被连续地控制,以一定的方向和速度跟随空间的路径。

8. 电源

一些机器人使用电动机,其他机器人使用液压执行机构。前者更快,后者在诸如喷漆等应用中更强大和有利,喷漆应用中,火花可能引起爆炸。

9. 驱动

一些机器人通过齿轮将电动机连接到关节;其他机器人直接将电动机连接到关节(直接驱动)。使用齿轮可以产生可测量的"间隙",这是在一个轴上的自由运动。较小的机器人手臂经常采用高转速、低扭矩直流电动机,其通常需要较高的传动比;缺点是会导致背隙。在这种情况下,经常使用谐波驱动。

第3部分 技术参数

技术参数反映了机器人的适用范围和性能,包括轴数、自由度、有效负载、工作空间、最大速度、分辨率、精度、重复性,其他参数有控制方式、驱动方式、安装、电源容量、重量、环境参数。

1. 轴数

两个轴可以到达平面上的任意点,三个轴可以到达空间中的任意点。为了完全控制手臂末端(即手腕)的方向,需要三个以上的轴。有些设计(如 SCARA 机器人)因为成本、速度和精确度限制了运动的可能性。

2. 自由度(DOF)

机器人末端执行器可以移动的独立运动的数量。它由机械手的运动轴数定义,通常与轴数相同,如图 2-3 所示。机器人的自由度反映机器人动作的灵活性。自由度越多,机器人就越接近人手的动作机能,通用性越好;但是自由度越多,结构就越复杂,对机器人的整体要求就越高。因此,工业机器人的自由度是根据其用途设计的。

3. 有效负载

机器人能举起的重量。最大有效负载是在保持额定精度的情况下,机器人机械手在速度降低时能承载的重量。额定负载是在最大速度时并保持额定精度的情况下测量的。不同工业机器人的有效负载如表 2-1 所示。

4. 工作空间

即接触空间,定义了机器人能够达到的空间区域的三维空间范围,如图 2-4 所示。

5. 最大速度

所有关节在互补方向上同时以最大速度移动时,机器人尖端的最大复合速度。这个速度是理论上的最大值,不能用于估计特定应用的周期时间。

6. 分辨率

可由机器人的控制系统检测或控制的运动或距离的最小增量。任何接头的分辨率都是编码器每转脉冲和驱动比的函数,取决于刀具中心点和接头轴之间的距离。

7. 精度

机器人与指定位置的接近度。机器人试图达到的与实际达到的位置之间的差。绝对精度是机器人控制系统目标点与由机械臂实际实现的点之间的差,而重复性是机械手重复到达同一点时的周期变化。

8. 重复性

机器人回到编程位置的程度。也就是系统重复相同动作或施加同一控制信号时到达相同位置的能力。是当机器人尝试执行特定任务时,系统的循环错误。

第 4 部分 运动学与动力学

运动的研究可分为运动学和动力学。

机器人运动学

机器人运动学将几何应用于构成机器人系统结构的多自由度运动链的运动的研究中。对几何的重视意味着机器人的链接被建模为刚体,并且其关节被假设为提供纯粹的旋转或平移运动。

机器人运动学研究运动链的尺寸和连通性与机器人系统中每个链路的位置、速度和加速度之间的关系,以便计划和控制运动并计算执行器的力和扭矩。

机器人运动学的一个基本用途是构成机器人的运动链的运动学方程。

正向运动学使用机器人的运动学方程,并通过指定的关节参数计算末端执行器的位置,如图 2-4 所示。计算可实现末端执行器的指定位置的关节参数的反过程被称为逆向运动学,如图 2-5 所示。机器人的尺寸及其运动学方程定义了机器人可达到的空间的体积,即为其工作空间。

正向运动学指定关节参数并计算链的配置。对于串联机械臂,可通过将关节参数直接代入串行链的正向运动学方程中来实现。对于并联机械臂,需要一组多项式约束条件才能将关节参数代入运动学方程,以确定可能的末端执行器位置的集合。

逆向运动学指定末端执行器位置并计算相关联的关节角度。对于串联机械臂,需要从运动学方程中获得一组多项式的解,并且得出链的多个配置。对于并联机械臂,末端执行器位置的参数简化了运动学方程,得出关节参数的公式。

正向运动学是指已知相应的关节值,计算末端执行器的位置、方向、速度和加速度。逆向运动学是指相反的情况,在路径规划时,根据给定的末端执行器数值,计算所需的关节值。

正向动力学是指已知作用力,计算机器人的加速度。正向动力学被用于机器人的计算机仿真。

逆向动力学是指计算要产生规定的执行器加速度所需要的作用力。该信息可用于改进机器人的控制算法。

第 3 单元 工业机器人的类型

第 1 部分 笛卡尔坐标机器人

笛卡尔坐标机器人(也称为线性机器人)是一种结构简单的工业机器人,如图 3-1 所示。

笛卡尔坐标机器人是一种手臂有三个棱柱形关节的机器人,轴线与笛卡尔坐标系一致,三个主要控制轴是线性的(即它们以直线而不是旋转的方式移动),并且两两垂直。三个滑动关节对应于手腕的上下、进出、前后运动。除了其他优点之外,这种机械装置简化了机器人控制解决方案。两端有水平构件支撑的笛卡尔坐标机器人有时也称为龙门式机器人;就机械结构而言,它们类似龙门起重机,尽管后者通常不是机器人。龙门式机器人通常相当大。

笛卡尔坐标机器人的一个普遍应用是计算机数控机床(CNC 机床)和 3D 打印(见图 3-2)。最简单的应用是用于研磨和拉丝机,当工具被升降到表面时拉丝机的笔或刨具在 X-Y 平面平移,从而进行精确设计。拾放机(见图 3-3)和绘图仪也基于笛卡尔坐标机器人的原理。

第 2 部分 SCARA 机器人

SCARA 是选择顺应性装配机器手臂或者选择顺应性关节机器手臂的缩写。

1981 年,SCARA(见图 3-4)在山梨大学教授 Hiroshi Makino 的指导下研发。

属性

SCARA 系统在 X,Y 轴方向上具有顺应性,而在 Z 轴方向具有良好的刚度,此特性特别适用于装配工作,例如将圆头针插入圆孔,故 SCARA 系统大量用于装配印刷电路板和电子零部件。

SCARA 的另一个属性是具有与人类手臂相似的联合双链臂布局,因此常用术语"铰接"。这个功能允许手臂延伸到狭窄的区域,然后退缩或"折叠"出来。这对于将部件从一个单元传送到另一个单元或用于装载/卸载被封闭的处理站是有利的。

特性

SCARA 通常比同类笛卡尔机器人系统更快、更洁净。其单基座安装需要较小的占地面积,并提供了一种简单、无阻碍的安装形式。但是,SCARA 可能比同类笛卡尔系统更昂贵,控制软件需要反向运动学进行线性插值运动。该软件通常和 SCARA 一起,且对终端用户是透明的。

大多数 SCARA 机器人都基于串行架构,这意味着第一台电动机应该携带其他所有电动机。还存在所谓的双臂 SCARA 机器人结构,其中两个电动机固定在基座上。第一台这样的机器人由三菱电机商业化推广。双臂 SCARA 机器人的另一个例子是 Mecademic 的 DexTAR 教育机器人。在国内,哈工大机器人集团推出了 HRG-HD1B5B 型五杆机器人,如图 3-5 所示。

第 3 部分　六轴关节型机器人

工业机器人具有各种轴配置。关节型机器人是具有旋转关节的机器人（例如，腿式机器人或工业机器人）。关节型机器人可以在简单的双联结构到具有 10 个或更多相互作用的接头的系统之间变动。它们可通过各种手段供电，包括电动机。

六轴关节型机器人是一类广泛使用的机械设备。绝大多数的关节型机器人具有 6 个轴，也称为六自由度。相比轴数较少的机器人，六轴机器人允许更大的灵活性，并且可以执行更广泛的应用。

轴 1——该轴位于机器人底座，允许机器人从左到右旋转。这种运动使工作区延伸至包括手臂任一侧和后面的区域。该轴允许机器人从中心点旋转达 180°的范围。该轴也称为 J1。

轴 2——该轴允许机器人的下臂向前和向后延伸。它是为整个下臂运动提供动力的轴。该轴也称为 J2。

轴 3——该轴延伸了机器人的垂直范围。它允许上臂升高和降低。在一些关节模型中，它允许上臂到达身体后面，进一步扩大工作范围，该轴使上臂更好地进入部件。该轴也称为 J3。

轴 4——配合轴 5 一起工作，此轴有助于端部执行器的定位和对部件的操纵。它被称为腕带，使上臂在水平方向和垂直方向之间以圆周运动的形式旋转。该轴也称为 J4。

轴 5——该轴允许机器人手臂的手腕向上和向下倾斜。该轴负责俯仰和偏航运动。俯仰或弯曲的运动是上下移动的，就像打开和关闭盒盖一样。偏航指向左或向右移动，像铰链门。该轴也称为 J5。

轴 6——这是机器人手臂。它负责扭转运动，允许其以圆周运动自由旋转，既可以定位末端执行器，也可以操纵零件。它通常能够在顺时针或逆时针方向上旋转超过 360°。该轴也称为 J6。

第 4 部分　码垛机器人

工业码垛是指将零件、箱子或其他物品装载到货板上或从货板卸载。自动码垛是指工业机器人码垛机自动执行应用。

码垛机器人（见图 3-7 和图 3-8）可以在许多行业中看到，包括食品加工、制造和运输行业。不同的机器人码垛机具有不同范围的有效负载和覆盖范围。

各种各样的手臂末端设计方式允许了不同的机器人码垛的灵活性。包夹持器包围住物品，并从底部支撑，而吸力和磁性夹具通常处理皱状物品，并从顶部抓住它们。使用码垛机器人自动化车间，可以提高装载和卸载过程的一致性。

码垛机器人（见图 3-9）于 20 世纪 80 年代初推出，并具有端部执行器，以抓住输送机或桌面上的产品，并将其定位到货板。

码垛机器人以低成本提供速度、重复性和精度。使用机器人码垛，货板以系统、整齐的方式堆垛和拆垛。

机器人码垛系统有很多优点。

(1) 安全。应用码垛机器人可以保护工人。机器人能够在冷藏和冷冻的环境中工作。

机器人码垛机可以处理对工人造成重复的背部伤害的任务。机器人提升重物,进一步加快生产。

(2) 灵活性。灵活的码垛机器人可以方便地更换产品或任务。它们可以快速重新编程,以适应各种产品和包装。可互换的 EOAT 选项允许机器人处理不同的材料或在应用程序之间来回切换。先进的视觉技术使所用的机器人能够区分部件和码垛的不同项目。

码垛机器人不仅可以提高生产线的速度,还能提高生产力。

第 5 部分　Delta 机器人

Delta 机器人是一种并联机器人。它由连接到基座通用关节的三根臂组成。关键设计特征是在臂中使用平行四边形,这保持了末端执行器的方向。相反,Stewart 平台可以改变其末端执行器的方向。

典型产品

1999 年,ABB Flexible Automation 开始销售其 Delta 机器人 FlexPicker,如图 3-9 所示。

2009 年,FANUC 发布了最新版的 Delta 机器人——FANUC M-1iA 机器人,并且随后为更大的有效负载发布该 Delta 机器人的变化版。FANUC 在 2010 年发布了 M-3iA(见图 3-10),用于更大的有效负载。最近,在 2012 年发布了 FANUC M-2iA 机器人,用于中型有效负载。

设计原则

Delta 机器人是一种并联机器人,它由连接底座和末端执行器的多个运动链组成。这种机器人也可以视为四杆联动的空间泛化。

Delta 机器人的关键概念是平行四边形的使用,将末端平台的移动限制为纯平移,即在没有旋转的情况下仅在 X,Y 或 Z 方向上移动。

机器人的底座安装在工作区上方,所有执行器都位于其上。三个中间连接臂从基座延伸。这些臂的端部连接到一个小的三角形平台。输入链接的启动将会使三角形平台沿着 X,Y 或 Z 方向移动,可以通过带或不带减速(直接驱动)的线性或旋转执行器进行启动。

由于执行器都位于基座,机械臂可以由轻的复合材料制成。因此,Delta 机器人的运动部件具有较小的惯性。这允许了非常高的速度和加速度。将所有臂一起连接到末端执行器的做法,增加了机器人的刚度,但降低了其工作量。

应用

Delta 机器人在工厂中普遍用于抓取和包装,因为它们可以相当快,有些可每分钟执行多达 300 次。

利用 Delta 机器人高速特性的行业有包装行业、医药行业。由于其刚度较大,它也用于手术。其他应用包括在洁净室进行高精度的电子元器件装配作业。

第 4 单元　ABB 机器人

第 1 部分　ABB 与 ABB 机器人

ABB

ABB(ASEA Brown Boveri)是一家总部设在瑞士苏黎世的瑞士跨国公司,主要从事机器人、电力和自动化技术领域。ABB 由瑞典公司 Allmänna Svenska Elektriska Aktiebolaget (ASEA)和瑞士公司 Brown,Boveri&Cie(BBC)于 1988 年合并组成,如图 4-1 所示。

ABB 是世界上最大的工程公司之一,其核心业务是电力和自动化技术。

ABB 机器人

ABB 是世界领先的工业机器人和机器人系统制造商,在全球超过 100 个地区的 53 个国家开展业务。

除了机器人以外,ABB 还生产和制造机器人软件、外围设备、工艺设备、模块化制造单元,并提供服务,如焊接、物料搬运、小零件组装、喷涂、拾取、包装、码垛和机器管理。

ABB 机器人的主要市场包括汽车、塑料、金属制造、铸造、太阳能、消费电子、木材、机床、制药及食品和饮料行业。强大的解决方案帮助制造商提高生产率、产品质量和工人安全。目前 ABB 已经在全球装设了超过 25 万台机器人。

ABB 机器人在我们的日常生活中发挥着重要作用,每时每刻我们都在使用着由 ABB 机器人制造或处理的产品。例如,ABB 机器人为雀巢、联合利华和吉百利等公司拾取、包装及码垛食品和饮料;它们帮宜家和 Tarkett 这两家最出名的品牌雕刻、砂磨、涂漆及包装家具和地板;它们为世界领先的品牌和制造商——苹果、戴尔、诺基亚等焊接、研磨、抛光和涂漆电脑、笔记本电脑、iPod、手机、相机和游戏机,如图 4-2 所示。

事实上,ABB 机器人不仅可以提高工业生产率,还可以大幅度提高能效和减少温室气体排放。例如,FlexPainter IRB 5500 喷涂机器人将世界各地的汽车工厂的车间能耗降低了 50%。

第 2 部分　ABB 产品系列

自 1988 年创立以来,为了满足市场的需求,ABB 公司推出了一系列的工业机器人产品,如搬运机器人、焊接机器人、装配机器人、喷漆机器人等。以下是 ABB 机器人主要型号的简介(具体的参数规格以 ABB 官方公布的最新数据为准)。

IRB 1410

IRB 1410 采用优化设计,专为弧焊而优化,额定载荷为 5 kg,工作范围达 1 440 mm。IRB 1410 提供电弧焊功能包,可以通过示教器来操纵,被广泛用于弧焊、材料处理和过程应用,如图 4-3 所示。

IRB 2400

IRB 2400 具有不同的版本和最佳精度,在材料处理、机器管理和工艺应用方面具有卓越的性能。IRB 2400 为产品生产提高效率,缩短交货时间,更快地交货。IRB 2400 是世界上最受欢迎的工业机器人之一，可以最大限度地提高电弧焊、加工、装卸等应用的效益，如图 4-4 所示。

IRB 52

IRB 52 是一款紧凑型喷涂机器人，广泛用于各行业易损件的喷涂。具有体形小巧、工作范围宽等特点，柔性与通用性俱佳，且其操作速度快、精度高、周期时间短。它包括了 ABB 独特的 IPS 集成工艺系统，以确保高品质、高精度的工艺过程和最终优质的涂层，以及降低涂料损耗，如图 4-5 所示。

IRB 360

近十五年来，ABB IRB 360 FlexPicker 先进的高速拾料和包装技术一直处于领先地位。与传统刚性自动化技术相比，IRB 360 具有灵活性高、占地面积小、精度高和负载大等优势。

IRB 360 系列现包括负载为 1 kg、3 kg、6 kg 和 8 kg 以及横向活动范围为 800 mm、1 130 mm 和 1 600 mm 等几个型号，这意味着 IRB 360 几乎可以满足任何需求。IRB 360 具有运动性能佳、节拍时间短、精度高等优势，能够在狭窄或者广阔空间内高速运行，误差极小。每款 FlexPicker 的法兰工具经过重新设计，能够安装更大夹具，从而高速、高效地处理分度带上的流水线包装产品，如图 4-6 所示。

IRB 910SC

通过设计选择顺应性装配机械臂（SCARA）、IRB 910SC，ABB 推出了能够在有限空间内运行的单臂机器人。ABB 的 SCARA 非常适合小零件装配、物料搬运和零件检查，如图 4-7 所示。

ABB 的 SCARA 机器人系列为各种通用应用而设计，如托盘包装、元件贴装、机器装卸和组装。这些应用要求快速、重复和准确的点对点运动，例如码垛、卸垛、机器装载/卸载和装配。ABB 的 SCARA 系列非常适合要求具有快速的循环时间、高精度、高可靠性的客户，用于小零件装配、实验室自动化和处方药分配。

IRB 14000（YuMi）

YuMi 双臂机器人于 2015 年 4 月 13 日正式推出市场，使人机协作成为现实。YuMi 是集柔性机械手、进料系统、基于相机的工件定位系统及尖端运动控制系统于一体的协作型小件装配双臂机器人解决方案。YuMi 是人机合作的愿景，如图 4-8 所示。

第 3 部分　ABB 机器人的结构

这里重点介绍 ABB 最新六轴工业机器人中的典型产品 IRB 120 机器人，如图 4-9 所示。

ABB IRB 120 机器人一般由三部分组成：机械臂、IRC5 控制器和示教器。

机械臂

机械臂又称操作机，是工业机器人的机械主体，用来完成规定任务的执行机构。主要由机械臂、驱动装置、传动装置和内部传感器组成。如图 4-10 所示，对六轴机器人而言，其机械臂主要包括基座、腰部、手臂（大臂和小臂）和手腕。

控制器

IRC5 控制器包含移动和控制机器人的所有必要功能。控制器包含两个模块，control module 和 drive module。两个模块通常合并在一个控制器机柜中。

control module 包含所有的电子控制装置，例如主机、I/O 电路板和闪存。control module 运行操作机器人所需的所有软件（即 RobotWare 系统）。

drive module 包含为机器人电动机供电的所有电源电子设备。IRC5 drive module 最多可包含 9 个驱动单元，它能处理 6 根内轴以及 2 根普通轴或附加轴，具体取决于机器人的型号。

使用一个控制器运行多个机器人时（MultiMove 选项），必须为每个附加的机器人添加额外的 drive module，但只需使用一个 control module。

示教器

示教器（FlexPendant）是一种手持式操作装置，用于执行与操作机器人系统有关的许多任务，例如运行程序、控制机器人本体、修改机器人程序。

示教器的外形结构如图 4-11 所示。

FlexPendant 由硬件和软件组成，其本身就是一台完整的计算机，是 IRC5 的一部分，通过集成线缆和接头连接到控制器。但是，"hot plug"按钮选项使其能在自动模式下断开 FlexPendant 的连接并继续运行。

第 4 部分　典型机器人——IRB 120

这里我们重点介绍 ABB 最新六轴工业机器人中的典型产品 IRB 120 机器人。

IRB 120

ABB 迄今最小的多用途机器人 IRB 120 仅重 25 kg，负载 3 kg（垂直腕为 4 kg），工作范围达 580 mm，是具有低投资、高产出优势的经济、可靠之选。它已经获得了 IPA 机构"ISO 5 级洁净室（100 级）"的达标认证，能够在严苛的洁净室环境中充分发挥优势。

IRB 120 专为基于机器人的柔性自动化的制造行业（如 3C 行业）而设计，通常用于装配、材料搬运等，如图 4-12 所示。

IRB 120 机器人的规格和特性见表 4-1。

表 4-1　机器人的规格和特性

规　格			
型号	工作范围	额定负荷	手臂负载※
IRB 120	580 mm	3 kg	0.3 kg
特　性			
集成信号接口	手腕设 10 路信号		
集成气路接口	手腕设 4 路气路（5 bar）		
重复定位精度	±0.01 mm		
机器人安装	任意角度		
防护等级	IP30		
控制器	IRC5 紧凑型		

※手臂负载是指小臂上安装设备的最大总质量，即 IRB 120 机器人小臂安装总质量不能超过 0.3 kg。

机器人的运动范围及性能见表 4-2。

表 4-2 机器人的运动范围及性能

运动		
轴运动	工作范围	最大速度/[(°)·s^{-1}]
轴 1 旋转	−165°～+165°	250
轴 2 手臂	−110°～+110°	250
轴 3 手臂	−90°～+70°	250
轴 4 手腕	−160°～+160°	320
轴 5 弯曲	−120°～+120°	320
轴 6 翻转	−400°～+400°	420
性能		
1 kg 拾料节拍		
25 mm×300 mm×25 mm※		0.58 s
TCP 最大速度		6.2 m/s
TCP 最大加速度		28 m/s^2
加速时间(0～1 m/s)		0.07 s

※(1) $s_1=25$ mm,$s_2=300$ mm,$s_3=25$ mm。
(2) 机器人末端安装 1 kg 物料,沿着 A→B→C→B→A 运行一周的时间为 0.58 s,如图 4-13 所示。

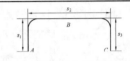

图 4-13 机器人运行轨迹图

工业机器人技能考核实训台(HRG-HD1XKA)

HRG-HD1XKA 型工业机器人技能考核实训台(专业版)(见图 4-14),是一款通用型六轴机器人实训台,可以选用任一品牌的紧凑型六轴机器人,结合丰富的周边自动化机构,配合标准化的工业应用教学模块,可通过上位机平台进行工业机器人虚拟仿真及 PLC、机器人编程等教学,主要用于工业机器人技术人才的培养教学和技能考核。

基础模块

该模块的示教盘上包含圆形槽、方形槽、正六边形槽、三角形槽、样条曲线槽以及 *XOY* 坐标系,如图 4-15 所示。使机器人和夹具沿前述各特征形状移动,可以进行简单轨迹示教的训练。该模块可以进行工具、工件坐标系标定,用于直线示教、圆弧示教、曲线示教学习。

第 5 单元　库卡机器人

第 5 部分　库卡与库卡机器人

库卡

KUKA 是一家从事工业机器人和提供工厂自动化解决方案的德国制造商。它由

Johann Joseph Keller 和 Jakob Knappich 于 1898 年在奥格斯堡创立。KUKA 名称由"Keller und Knappich Augsburg"中所用词的第一个字母缩写而成。100 多年以来,KUKA 始终秉承使其在世界上取得成功的理念和创新精神。全世界的 12 000 多名 KUKA 职员正在开发面向未来的智能机器人自动化解决方案——创新、综合、高效。它被称为"橙色智能"。而连接现实和虚拟世界的"橙色智能"是 KUKA 创新、综合和高效的动力源泉。

KUKA 如今成为自动化解决方案的世界领先供应商之一。它代表着自动化领域的创新,是工业 4.0 的推动者。

库卡机器人

作为机器人技术与自动化技术领域的专家,KUKA 机器人有限公司是世界领先的工业机器人制造商之一。KUKA 机器人种类齐全,几乎涵盖了所有负载范围和类型,并确立了人机协作(MRK)领域的标准。

1. 工业机器人领域的先驱

KUKA 在工业机器人制造方面已经有 40 多年历史。早在 20 世纪 70 年代就已奠定了今天的工业 4.0 基础:第一台工业机器人 FAMULUS。1996 年,KUKA 机器人有限公司的工业机器人开发取得质的飞跃:当时,由 KUKA 公司开发的首个基于 PC 的控制系统已投放市场。由此开创了"真正的"机电一体化时代:以软件、控制系统和机械设备的完美结合为特征。今天,KUKA 机器人有限公司为各行各业提供种类繁多的产品和量身定制的自动化解决方案。KUKA 工业机器人能够提高产品质量,减少昂贵材料和有限资源的使用。

KUKA 机器人有限公司的愿景是使工业机器人成为人类在生产中的智能化助手:人与机器人携手合作,通过能力互补成为理想的工作伙伴。

2. 产品和行业

KUKA 机器人有限公司为客户提供量身定制的工业机器人解决方案。从机器人部件到控制系统,再到合适的软件,各个行业的客户都可以从创新的技术和精心的设计中获益。

KUKA 机器人的产品系列有:几乎涵盖所有作业范围和负载能力的六轴机器人,耐高温、防尘及防水的机器人,应用于食品和制药行业的机器人,净化室机器人,码垛机器人,焊接机器人,冲压连线机器人,架装式机器人和高精度机器人。KUKA 机器人有限公司的解决方案主要的应用行业和客户有:气体保护焊及其他众多焊接工艺、机床行业、铸造业和锻造业、塑料工业、电子设备工业、食品工业、汽车配件供应商和汽车制造商(见图 5-1)。

第 2 部分　库卡机器人产品系列

KUKA 提供全方位的工业机器人,无论多么具有挑战性的应用,顾客总能找到合适的机器人产品。以下是 KUKA 机器人主要型号的简介(具体的参数规格以 KUKA 官方公布的最新数据为准)。

KR 16

KR 16 是 KUKA 应用最广泛的六轴工业机器人之一,如图 5-2 所示。它扩展方便,可用于不同的组合。它的关节运动系统使其成为完成低负载区所有点和轨迹控制任务的理想伙伴。

KR 700 PA

KR 700 PA 是非常适用于更高负载的码垛机器人,如图 5-3 所示。它的工作范围非常大,结构紧凑,非常适用于各种应用情况。无论用在哪里,都能确保最佳的循环时间。

KR AGILUS sixx

KR AGILUS sixx 是 KUKA 按照最高工作速度设计的紧凑式六臂机器人,如图 5-4 所示。这种小型机器人有不同的款式、安装位置、工作范围和负载能力,是应用广泛的高精度"艺术家"。无论是在地面、天花板或墙壁,由于使用内置的能源供应系统和可靠的 KR C4 compact 控制系统,它都可以在最小空间内实现最高的精度。

KR CYBERTECH ARC nano

KR CYBERTECH ARC nano 产品系列是连续轨迹应用(如气体焊接、涂胶和涂刷密封剂)的最佳选择,如图 5-5 所示。这款工业机器人具有最佳的性能和最高的功率密度,能够以最低的成本实现最大的经济效益。其 50 mm 空心轴腕部是面向未来的创新设计,空心轴可以减少基轴运动,同时具有极短的周期时间和最高的运动精度。

KR QUANTEC pro

KR QUANTEC pro 针对 90~120 kg 的高负载区域进行了优化,如图 5-6 所示。该款机器人结构紧凑、性能强大且极为精准——为工业 4.0 开辟了创新的单元设计方案。其灵巧的手部和更小的干扰轮廓确保了最高的精确度和速度。因此,它非常适合于点焊、焊接和搬运任务。

LBR iiwa

LBR iiwa 是第一款量产的灵敏型机器人,如图 5-7 所示,也是具有人机协作能力的机器人。它为灵敏型工业机器人技术开创了新的纪元,也为创新、可持续发展的生产流程奠定了基础。它首次实现了人类与机器人之间的直接合作,以完成高灵敏度需求的任务。它也形成了新应用的可能性,可以提高经济效益并且达到最高效率。LBR iiwa 有两种机型可供选择,负载能力分别为 7 和 14 kg。

第 3 部分　库卡机器人的结构组成

本节重点介绍高负载产品 KR QUANTEC pro,它是一款 KUKA 六轴工业机器人,如图 5-8 所示。KR QUANTEC pro 机器人一般由三部分组成:操作机、KR C4 控制器和示教器。

操作机

操作机又称机器人本体,是工业机器人的机械主体,用来完成规定任务的执行机构。主要由机械臂、驱动装置、传动装置和内部传感器组成。对六轴机器人而言,其机械臂主要包括基座、腰部、手臂(大臂和小臂)和手腕,如图 5-9 所示。

控制器

KUKA KR C4 控制系统是现在和未来的自动化先锋。它可降低集成、保养和维护方面的费用。同时还将持续提高系统的效率和灵活性——由于通用的开放式工业标准。KR C4 控制器在软件架构中集成了 Robot Control、PLC Control、Motion Control(如 KUKA.CNC)和 Safety Control。所有控制系统都共享一个数据库和基础设施。为最理想地集成到自动化环境,KRC4 有五种规格可供选择:KR C4 compact,KR C4 smallsize-2,KR C4,KR C4 midsize 和 KR C4 extended。同时,这些规格都考虑了堆叠放置以及防尘、防潮和其他防护措施方面的要求。

KR C4 由以下部件构成,如图 5-10 所示。

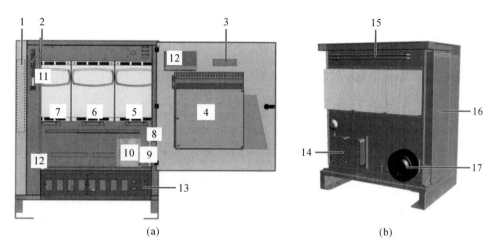

图 5-10　KR C4 控制器组成
(a) 正视图　(b) 后视图

1—电源滤波器；2—主电源；3—控制器系统面板(CSP)；4—控制计算机(KPC)；
5—驱动电源(驱动控制器 7 和 8 轴，可选)；6—伺服驱动器(轴 4～6)；7—伺服驱动器(轴 1～3)；8—制动过滤器；
9—内控制单元(CCU)；10—安全接口面板(SIB)/SIB 扩展；11—瞬态抑制器；12—电池；13—连接板；
14—低压供电单元；15—制动电阻；16—热交换器；17—外部风扇

示教器

smartPAD 是一种手持式操作装置，KUKA 机器人的绝大部分操作均可以通过它来完成。

KUKA smartPAD 是为了以简单方式完成复杂操作任务而研发的。它可以广泛应用，即使是未经练习的用户也能轻松操作，能够操作所有配备 KR C4 控制系统的 KUKA 机器人。

示教器的外形结构如图 5-11 所示。

smartPAD 由硬件和软件组成，它本身就是一台完整的计算机，通过集成线缆和接头与控制器相连接。

第 4 部分　典型机器人——KR 6 R700

这里我们重点介绍 KUKA 小型六轴工业机器人中的典型产品 KR 6 R700 sixx 机器人。

KR 6 R700 sixx

它是 KUKA 按照高工作速度设计的紧凑式六臂机器人，如图 5-12 所示。无论是在地面、天花板或墙壁，由于使用集成式能源供应系统和可靠的 KR C4 compact 控制系统，它都可以在最小空间内实现最高的精度。

它的特点如下。

(1) 最短的循环时间。KR 6 R700 有 6 个轴，专为极高的作业速度设计，同时还具有极高的精确度。

(2) 节省空间的集成。极小的占用空间以及可选择的安装方式(可安装于地面、天花板或墙壁)，使得它适应能力极高。

(3) 集成式能源供应系统。集成式能源供应系统被敷设在机器人内部，十分节省空间。

它包括 EtherCAT/以太网(总线电缆)、3 个二位五通阀(压缩空气)、直接气路以及输入和输出信号接口。

(4) KUKA.SafeOperation.机器人在安全性方面树立了标准。只有这种机器人可提供 KUKA.SafeOperation 功能,明显简化了人机协作的集成工作。

KR 6 R700 sixx 机器人的规格和特性见表 5-1。

表 5-1 KR 6 R700 sixx 机器人的规格和特性

规　　格		
型号	工作范围	额定负载
KR 6 R700 sixx	706.7 mm	6 kg
特　　性		
重复定位精度	±0.03 mm	
安装位置	地板、天花板、墙壁	
占地面积	209 mm × 207 mm	
质量(不包括控制器)	约 50 kg	
环境温度	+5～+45 ℃	
防护等级	IP 54	
控制器	KR C4 紧凑型	

KR 6 R700 sixx 机器人的运动范围及性能见表 5-2。

表 5-2 KR 6 R700 sixx 机器人的运动范围

运　　动		
轴	运动范围	最大速度
轴 1(A1)	−170°～+170°	250°/s
轴 2(A2)	−190°～+45°	250°/s
轴 3(A3)	−120°～+156°	250°/s
轴 4(A4)	−185°～+185°	320°/s
轴 5(A5)	−120°～+120°	320°/s
轴 6(A6)	−350°～+350°	420°/s

工业机器人技能考核实训台(HRG-HD1XKA)

HRG-HD1XKA 型工业机器人技能考核实训台(专业版),是一款通用型六轴机器人实训台,如图 5-13 所示。可以选用任一品牌的紧凑型六轴机器人,结合丰富的周边自动化机构,配合标准化的工业应用教学模块,可通过上位机平台进行工业机器人虚拟仿真及 PLC、机器人编程等教学,主要用于工业机器人技术人才的培养教学和技能考核。

激光雕刻模块

激光器沿着雕刻面板的轨迹运行,用于激光雕刻应用的学习,以达到基础功能熟练应用及 I/O 信号配置的目的,如图 5-14 所示。

第6单元　安川机器人

第1部分　安川与安川机器人

安川

安川电机是世界上最大的交流变频器驱动、伺服运动控制和机器人自动化系统制造商。安川电机成立于1915年，总部在日本。19世纪60年代末，安川电机引领世界，提出了"机电一体化"的概念。安川通过结合用户设备及其电气产品来创造卓越的品质和功能，使得这个概念得以提升。

自1915年以来，安川就一直通过自动化产品改善全球生产力而服务世界。在2016年，安川开始中期商业计划"冲刺25"，这是实现长期商业计划"2025规划"的第一步。

安川机器人

自从1977年全电动工业机器人"MOTOMAN"在日本首次亮相后，安川机器人成为世界工业机器人市场的主要角色。从汽车生产的弧焊作业开始，安川机器人一直是市场领导者，它对每一个工业领域都起到积极作用，如焊接、包装、装配、涂料、切割、物料搬运和自动化生产，以及在洁净室对液晶、有机EL显示器的处理和运输及半导体制造，如图6-1所示。

作为一个近年来解决劳动力短缺的方法，安川机器人的应用范围正向几乎没有应用过机器人自动化的食品行业和其他市场扩展。安川推动了机器人新市场的创造，加速了开放创新。

近期，安川机器人通过提升速度、提高准确性、应对复杂运动和加强与人类共存的安全功能，来扩大在工业和其他领域的影响力和表现。在未来，安川机器人将持续推进关于积极参与贴近人类应用领域的发展。

第2部分　安川机器人产品系列

自1977年以来，安川已经相继商业化并销售了各种用途的优质机器人，用途以电弧焊为中心——其专业领域之一，包括点焊、处理、装配、涂装、液晶面板的转移、半导体晶圆的转移等。以下是安川机器人的主要型号的简要说明（具体的规格参数以安川官方公布的最新数据为准）。

MH5SⅡ

MH5SⅡ是安川一款紧凑型高速六轴机器人，它在包装、物料搬运、机器管理和分配等要求多功能的应用中都提供了优越的性能，如图6-2所示。

MH5SⅡ工作范围为706 mm，在同类中工作范围最大。全轴低功率输出，无须设置安全性防护，可以有效减少机器人与周边设备的碰撞。它采用DX200控制器，该控制器可以控制多达8个机器人。由于该机器人的紧凑型设计和内置的碰撞检测功能，多个机器人可以很容易地适用于同一个生产设备。

MPP3S

四轴高速并联机器人MPP3S结合了速度增量设计，具有高负载能力，工作范围大，如图

6-3 所示,适用于高速、高精度的码垛,取件和包装。

MPP3S 可以安装于一个狭小的空间。它还能够在大范围内进行高速搬运,从而提高生产力。除了高速的机器人运动之外,较短的吸入时间也有助于减少循环周期。机器人的空心结构(工业中第一次采用)使得空气阀能够安装于并联机械臂的内部。此结构大幅缩短了从动臂管的长度,减少了循环周期。MPP3S 也是为了保持高度清洁的食品搬运而设计的。

MPL300 II

多功能、强大的四轴机器人 MPL300 II 在包装、箱体码垛和大多数生产线的尾部作业或配送中心自动化方面提供了高性能,如图 6-4 所示。MPL300 II 的垂直伸展长度为 3 024 mm,水平伸展长度为 3 159 mm,这使得它具有高码垛负载(有效负载为 300 kg)。从基座到机械臂末端工具的内置路径航线和电缆最大化了系统的可靠性。

EPX2800

灵活、高性能的 MOTOMAN 喷涂机器人 EPX2800 提高了喷涂的质量、一致性和产量,同时大幅降低了运营成本,减少了材料的浪费。EPX2800 机器人是汽车和其他工业涂料应用的理想之选,如图 6-5 所示。它提供了卓越的性能,在喷涂和分配应用上有光滑、一致性高的出色效果。它采用中空手腕设计,是喷涂波状外形零件(如内部/外部表面)的理想之选。它非常适合装有喷雾装置的涂抹器,可避免软管和零部件之间的干扰从而确保最佳循环周期和机器人动作。

MYS450F

MYS450F 是一款四轴 SCARA 机器人,速度快,结构紧凑,只需要很小的安装空间,如图 6-6 所示。MYS450F 在装配、零部件搬运、包装和实验室自动化等应用方面具有优越的性能。SCARA 机器人很容易就可以与现有机器人的应用进行集成,从而扩展目前自动化过程。它是需要具有抓取和定位能力的大型多进程系统的理想之选。

SIA10D

SIA10D 是一款紧凑和强大的七轴单臂机器人,适用于自动化作业,如装配、检验、机器管理和搬运,如图 6-7 所示。SIA10D 采用高手腕性能和完全集成式供应电缆的革命性设计,使其在密闭空间里能够以惊人的自由运动进行工作。SIA10D 具有最小轨迹和高运动灵活性,它可以定位在正常工作区域(如地板、天花板、墙壁、斜坡或机械安装)而不会限制任何轴的运动范围。为了节省宝贵空间,SIA10D 可以安装在机器之间,这也为机器维护、调试或测试提供了开放的空间。

第 3 部分　安川机器人的结构组成

本节重点介绍一款高性能机器人产品——MH12,它是一款安川六轴工业机器人,如图 6-8 所示。MH12 机器人一般由三部分组成:操作机、DX200 控制器和示教器。

操作机

操作机又称机器人本体,是工业机器人的机械主体,用来完成规定任务的执行机构。主要由机械臂、驱动装置、传动装置和内部传感器组成。对六轴机器人而言,其机械臂主要包括基座、腰部、手臂(大臂和小臂)和手腕,如图 6-9 所示。

控制器

安川新型控制器 DX200 以强大的 PC 结构和机器人工作单元的系统级控制为特点。

该控制器使用了多个机器人控制技术专利,如 I/O 设备和通信协议,而且它提供了内置梯形图逻辑处理,包括 4 096 个 I/O 地址、各种现场总线网络连接、高速的电子服务器连接、接口面板,用以在示教器上显示人机界面(HMI)。DX200 通常不需要单独的 PLC 和 HMI,并且在系统层面上节约成本,同时降低了工作单元的复杂性,提高了系统整体的可靠性。动态干涉区域保护了机器臂,提供了高级碰撞避免功能。先进的机器人运动(ARM)控制提供了高性能、最佳的路径规划,大大减少示教时间。它支持多台机器人或其他设备的协调运动。小巧、轻便的 Windows® CE 编程器具有有多窗口显示功能的彩色触摸屏。编程特性的设计目的是使用最少的击键次数,并通过新的函数包和超过 120 个函数来实现。此外它根据应用程序和机器人的大小能够节约 38%~70% 的能耗。它配有可选的 3 级功能安全单元(FSU),并允许建立 32 个安全单元和 16 个工具。

其主要特点如下:
(1) 包括 120 多个函数的应用程序特定函数包;
(2) 可选的 3 级功能安全单元(FSU);
(3) 生产效率高;
(4) 集成成本低;
(5) 集成单元控制性能;
(6) 可靠性和能源利用率高;
(7) 易维护;
(8) 编程简单;
(9) 便携、紧凑的闪存槽和 USB 接口,方便储存备份。

示教器

示教器是一种手持式操作装置,安川机器人的绝大部分操作均可以通过它来完成。它的主要作用是进行机器人运动示教。

示教器的外形结构如图 6-10 所示。

示教器是由硬件和软件组成,它本身就是一台完整的计算机,通过集成线缆和接头与控制器相连。

第 4 部分　典型机器人——MH12

这里我们重点介绍安川高速六轴工业机器人中的典型机器人产品 MH12。

MH12

灵活、高速六轴机器人 MOTOMAN MH12 是安川 MOTOMAN MH 系列产品线上新增的一员,它的负载为 12 kg。

通过同级别最高速性能的实现,安川致力于生产力的发展。上臂采用空心轴结构,通过在中空手臂中安置电缆,降低由于操作限制而导致的电缆干扰。因此,像简化机器人示教操作或消除电缆断裂问题之类的问题的可维护性得以改进。新型弧焊机器人采用的全方位结构也应用于该款多目标应用机器人。这有助于减少夹具和工件之间的干扰区域。优越的性能可以应用到搬运操作,例如适应大型工件旋转的需要。包括机器人的手臂在内,伺服浮动功能能够减少诸如注塑机中由活塞往复推动引起的工件的搬运。

MH12 机器人的规格和特性见表 6-1。

表 6-1　MH12 机器人的规格和特性

规　格		
型号	工作范围	额定负载
MH12	1 440 mm	12 kg
特　性		
重复定位精度	±0.08 mm	
安装温度	0～+45℃	
安装湿度	20%～80%	
质量	130 kg	
电源容量	1.5 kVA	

MH12 机器人的运动范围见表 6-2。

表 6-2　MH12 机器人的运动范围

运　动		
轴	运动范围	最大速度
S(轴1)	-170°～+170°	220°/s
L(轴2)	-90°～+155°	200°/s
U(轴3)	-175°～+240°	220°/s
R(轴4)	-180°～+180°	410°/s
B(轴5)	-135°～+135°	410°/s
T(轴6)	-360°～+360°	610°/s

工业机器人技能考核实训台(HRG-HD1XKD)

HRG-HD1XKD 型工业机器人技能考核实训台(标准版)如图 6-11 所示,集工业机器人的实训、考核于一体,是一款功能实用、模块丰富、可扩展的工业机器人教学装备,适用任一品牌的工业六轴机器人。集成 TCP 标定等基础教学模块,配套激光雕刻、工件搬运、装配等工业应用教学模块,便于由浅入深地学习工业机器人操作技能。该实训台也可选配机器人其他周边知识的教学模块,如视觉系统、PLC 编程系统等,非常适合于高职院校的工业机器人技术人才培养教学和技能考核。

焊接模块

模拟焊枪沿着需要焊接的点的位置形成焊接轨迹,为了很好地演示焊接功能,机器人需要在转角位置点处理好焊枪姿态变化,并在整个焊接过程中对速度和姿态进行控制,如图 6-12 所示。

第7单元　发那科机器人

第1部分　发那科与发那科机器人

发那科

自1956年以来，当FANUC首次在日本私企中成功开发了伺服机械，它就一直追求工厂的自动化。

发那科是一个集团公司，主要是日本发那科公司，提供自动化产品和服务，如机器人和计算机数控系统。发那科是世界上最大的工业机器人制造商之一。发那科是作为富士通的一部分开始的，早期开发数控（NC）和伺服系统。公司的名字是富士自动数控的缩写。

1972年，计算控制部门独立出来，发那科有限公司成立。该公司的客户包括美国和日本的汽车和电子产品制造商。FANUC在46个国家有超过240个合资企业、子公司和办事处。它是最大的数控机床制造商，占全球市场的65%的市场份额，也是工厂自动化系统的全球领先制造商。

发那科机器人

自1974年FANUC首台机器人问世以来，FANUC一直致力于机器人技术上的领先与创新，是世界上唯一一家由机器人来做机器人的公司，是世界上唯一一个提供集成机器人视觉系统的企业，是世界上唯一一家既提供智能机器人又提供智能机器的公司。

FANUC机器人产品系列多达240种，负载从0.5 kg到2.3 t，产品广泛应用在装配、搬运、焊接、铸造、喷涂、码垛等不同生产环节，满足客户的不同需求，如图7-1所示。

2008年6月，FANUC成为世界上第一个装机量突破20万台机器人的厂家；2011年，FANUC全球机器人装机量已超过25万台，市场份额稳居第一。

第2部分　发那科机器人产品系列

为了满足市场的需求，FANUC公司推出了一系列的工业机器人产品，如搬运机器人、焊接机器人、装配机器人、喷漆机器人等。以下是FANUC机器人主要型号的简介（具体的参数规格以FANUC官方公布的最新数据为准）。

M-2iA/3S

M-2iA/3S系列是适用于高速搬运和装配的中型并联机器人。M-2iA/3S是一款手腕采用一轴旋转结构的三轴机器人，其负载为3 kg，如图7-2所示。它适用于传送带上面需要方向调整的高速拾取作业。而且它采用了完全密封的结构（IP69K），从而可应对高压喷流清洗，致力于食品领域的自动化生产。通过和视觉传感器（iRVision）或力传感器配套使用，M-2iA/3S机器人可以使用最新的智能化功能。

M-410iB/160

M-410iB/160系列是为了实现物流系统的机器人化而开发的大型物流智能机器人。M-410iB/160机器人是负载为160 kg的高速搬运型机器人，如图7-3所示。客户可以使用最新的智能化功能 iRVision 和 ROBOGUIDE，例如检测堆垛高度、识别工件的大小和种类、视觉分拣和从货物托盘上卸货。

ARC Mate 120iC

ARC Mate 120iC 是电缆内置式弧焊机器人，负载为 20 kg，如图 7-4 所示。通过采用高刚度的手臂和最先进的伺服技术，提高了机器人的加减速性能，缩短了过渡动作时间，提高了生产率。手腕轴采用了独特的驱动机构，从而实现了苗条的电缆内置式手腕。由于实现了稳定的焊炬电缆管理，可以使用 ROBOGUIDE 进行脱机示教，以大幅度减少机器人的示教时间。

M-900iA/260L

M-900iA/260L 是一款重载、长臂型机器人，其负载为 260 kg，最大有效范围为 3.1 m，如图 7-5 所示。此款机器人最适合于远距离搬运大体积的工件，如搬运汽车的车身和将大型铸件安装到加工夹具上。机械手腕部具有相当于 IP67 的耐环境性（防尘、防滴），即使在恶劣环境下也可放心使用。

P-250iB

P-250iB 机器人是目前市场上最柔性化、最先进的喷涂机器人之一，如图 7-6 所示。P-250iB 拥有多种安装方式，以此来满足最为苛刻的喷涂应用环境。"Open Architecture"设计理念使得机器人既可以安装标准的 FANUC 集成式工艺设备，也可以简单方便地集成第三方工艺设备。可选配的手臂内置气动控制柜将安装在机器人的垂直臂上，内部可以安装 FANUC 认证的组件，使用这些组件可以提高整体性能和喷涂质量。

CR-35iA

CR-35iA 是可与人相互协调作业的负载为 35 kg 的协同作业机器人，如图 7-7 所示。它无需安全栅栏就可以与人共享某个区域进行作业，如重零件的搬运和零件的装配。接触到人时，机器人会安全地停止。给人以安心感的绿色软护罩可以缓和冲击力，防止人被夹住。CR-35iA 已经取得符合国际标准的 ISO 10218-1 的安全认证。它采用与以往的机器人相同的高可靠性设计。

第 3 部分　发那科机器人的结构组成

本节重点介绍一款 FANUC 典型机器人产品——LR Mate 200iD/4S 机器人，它是一款 FANUC 迷你型六轴工业机器人，如图 7-8 所示。LR Mate 200iD/4S 机器人一般由三部分组成：操作机、R-30iB Mate 控制器和示教器。

操作机

操作机又称机器人本体，是工业机器人的机械主体，用来完成规定任务的执行机构。主要由机械臂、驱动装置、传动装置和内部传感器组成。对六轴机器人而言，其机械臂主要包括基座、腰部、手臂（大臂和小臂）和手腕，如图 7-9 所示。

控制器

FANUC R-30iB Mate 控制器是一款高性能控制器，为生产车间带来了更高层次的生产力。它提供了硬件和最先进的网络通信、iRVision 集成和运动控制功能。此外，FANUC 减小了控制器空间，为制造商节省了车间空间，或允许制造商为多机器人设备堆叠控制器。

由于控制器与外部电源开关之间需要较少的能源消耗，因此该控制器还有利于节能。

它还有冷却风扇自动停机功能，减少休眠期间的功率转换，在机器人空闲时，制动控制功能通过自动制动电动机减少功率，ROBOGUIDE 功率优化功能为客户降低功率和优化节能。

该控制器也有一个可选的节能设计，在制动期间可以恢复动能并返回至系统中，以便在接下来的周期内重新被使用。

而控制器节能的增加，反过来，会降低能源成本，通过加强释放制动器、机器人运动优化和机器臂智能振动控制来减少周期时间和使得运动更平稳，从而提高了公司的生产率。

示教器

iPendant 是一种手持式操作装置，用于执行与操作机器人系统有关的几乎所有任务。

示教器的外形结构如图 7-10 所示。

iPendant 是一个高度可定制的、便携式显示和操作面板。iPendant 有一个 4D 图形触摸屏，可以显示处理信息和实际流程路径，允许其更方便地建立。如果有必要，可以添加一个触摸面板接口作为选配，所有的键可以在简易键表中进行定制化。另外，连接单元使 iPendant 具有可移植性或安全性。

第 4 部分　典型机器人——LR Mate 200iD/4S

这里重点介绍 FANUC 典型六轴工业机器人产品 LR Mate 200iD/4S。

LR Mate 200iD/4S

LR Mate 200iD/4S 是一款大小和人的手臂相近的迷你机器人。苗条的机械臂减少了在狭小空间内与周边设备的碰撞。其具有同级别机器人中的最轻的机构部分，能够容易地把它安装在加工机械内部或者进行吊装。另外，通过采用高刚度的机械臂和最尖端的伺服控制技术，实现了在高速作业中平滑运动而不晃动。LR Mate 200iD/4 具有手腕负载容量大的特点，可以轻松地对应需要搬运多个工件的作业。因为设备的传感器电缆、附加轴电缆、电磁阀、空气导管和 I/O 电缆都集成在机械臂中，所以容易实现手工布线。

LR Mate 200iD/4S 机器人的特性见表 7-1。

表 7-1　LR Mate 200iD/4S 机器人的特性

	特　　性
型号	LR Mate 200iD/4S
工作范围	550 mm
手腕最大负载	4 kg
安装方式	地面、顶吊、倾斜角
重复定位精度	±0.02 mm
质量	20 kg
输入电源容量	1.2 kVA

LR Mate 200iD/4S 机器人的运动范围见表 7-2。

表 7-2　LR Mate 200iD/4S 机器人的运动范围

轴	运动	
	运动范围	最大速度
J1（轴1）	340°	460°/s
J2（轴2）	230°	460°/s
J3（轴3）	402°	520°/s
J4（轴4）	380°	560°/s
J5（轴5）	240°	560°/s
J6（轴6）	720°	900°/s

HRG-HD1XKB

HRG-HD1XKB 型工业机器人技能考核实训台（标准版），如图 7-11 所示，是一款通用型六轴机器人实训台，可以适用任一品牌的紧凑型六轴机器人。结合丰富的周边自动化机构，配合标准化的工业应用教学模块，其主要用于工业机器人技术人才的培训教学和技能考核。

异步输送带模块

输送带运行后，首先将工件放入输送带上，工件沿输送带运行至末端；然后，末端光电开关感应到物料并反馈给系统，输送带停止；最后，机器人移动至输送带末端并抓取工件将其放置于物料托盘上。以上实现了生产线流水作业仓储功能演示，如图 7-12 所示。

第8单元　水平关节机器人

第1部分　爱普生水平关节机器人

爱普生是全球 SCARA 机器人领先制造商。它在 1981 年开始开发机器人。在过去的 30 年里，爱普生用它的机器人专业技术制造出许多成功的爱普生机器人，这些机器人以精确、速度和可靠性著称。

以下是爱普生 SCARA 机器人主要系列的简介（具体的参数规格以爱普生官方最新公布的数据为准）。

G 系列

利用爱普生独特智能运动控制技术设计的 G 系列空间节约型机器人，提供了行业最好的速度和准确性，如图 8-1 所示。

爱普生加强了对用户重要的规范和功能来设计和开发系统，为了多产品的生产优化了系统，有的支持水和油雾环境。

爱普生 G 系列 SCARA 机器人的主要特点如下：

（1）快速、准确的运动缩短了工作时间；

（2）精确和光滑的直线和圆弧运动提高了复杂工作的质量；

（3）内部布线的无管设计使得运动范围增加了 20%；

（4）机器人可以选择安装在平面、墙壁或天花板上，以达到最好的布局和最高的空间利用率。

RS 系列

RS 系列机器人已不仅仅是 SCARA 机器人，我们称它们为 SCARA＋，如图 8-2 所示。与普通的 SCARA 机器人不同，爱普生 RS 系列具备一种独特的能力——能够移动第一轴下面的第二轴，因而使其能够在整个工作区内移动。普通的 SCARA 机器人必须绕自己移动，因而在中间留下了一个无法触及的大区域，因此爱普生 RS SCARA＋机器人在工作范围和速度上具有独特优势。爱普生 RS 系列 SCARA＋机器人的重复定位精度低至 0.010 mm，循环时间缩短至 0.339 s。

RS 系列简直是零占地面积机器人。它们可以很容易地集成到紧凑的装配单元中。与其他"旧"SCARA 设计相比，RS 系列机器人独特的工作空间设计改善了循环时间。这意味着可在更短的时间内加工更多的部件，同时使用一小部分的地面空间，带来更多的利润。

LS 系列

LS 系列 SCARA 机器人的设计目的是为了实现有效的成本控制，专门面向那些希望在不影响产品性能的情况下获取更大价值的厂商，如图 8-3 所示。但是和所有的爱普生机器人一样，它们仍然具有相同的性能和可靠性。此外，通过卓越的工作空间运用度，爱普生 LS 系列机器人提供占用空间小的解决方案，能够满足当今工厂的空间需求。出色的循环周期以及工业领先的易用性和可靠性，使它们成为高速应用的首选机器人，且性价比高。

T 系列

T 系列紧凑 SCARA 机器人的特点是内置了控制器，在设置和维护过程中无需处理复杂的布线问题，且无电池运动单元提高了成本效率，有助于降低总运营成本，如图 8-4 所示。

T 系列在接近末端执行器的地方提供了一个 I/O 通信端口。这个端口使电缆连接到末端执行器更容易，并为它提供电源，不再需要将长电缆布线到控制器。电缆管道，包括气动软管和电缆，比以前的型号要短。较短的设计使机器人在移动时保持稳定，使得在管道外的电缆走线更加容易。

第 2 部分 雅马哈水平关节机器人

YAMAHA 是全球 SCARA 机器人领先制造商之一。YAMAHA 机器人从水平多关节机器人的生产起步。自 1979 年生产了最初的水平多关节机器人"CAME"以来，YAMAHA 30 年来坚持不懈地从事水平多关节机器人的开发。

其 YK 系列是行业内顶级丰富的产品系列，完全无皮带式的结构发挥出水平多关节机器人特征的极限。以下是 YK 系列机器人主要型号的简介（具体的参数规格以 YAMAHA 官方公布的最新数据为准）。

YK350TW

YK350TW 是全方位型 SCARA 机器人，其工作范围为 350 mm，如图 8-5 所示。它克服了以前 SCARA 机器人和并联机器人的缺点，兼具高定位精度与高速性能。YK350TW 采用悬挂结构和大范围的机械臂旋转角度，可以覆盖机器人下方 1 000 mm 的全区域。

YK150XG

YK150XG 是一款微型 SCARA 机器人，其有效负载为 1 kg，如图 8-6 所示。以同等级中唯一采用完全无皮带式结构的超小型机型实现顶级的高刚度、高精度。通过采用 ZR 轴直接耦合实现完全无皮带式结构,这种直接驱动结构大大缩短了节拍时间。另外,YK150XG 通过改善减速比和电动机最高转速,实现了最高速度的大幅度提高。YK150XG 从前面和上面都可以打开外罩,其外罩和电缆为独立结构,容易维护。

YK350XG

小型 SCARA 机器人 YK350XG 的额定负载为 5 kg,适用于小型零件的组装、搬运、压入作业与涂胶等作业,如图 8-7 所示。它的前端旋转轴直接连接减速器。与一般减速后使用皮带的传输结构相比,它由于具有极高的 R 轴容许惯性力矩,因此偏移的工件也可以高速运作。通过改变电缆布局,使电缆高度低于主机外罩,并且,通过使用型材底座和低总高电动机,实现了同级别产品中的最低总高度。

YK550XG

中型机型 YK550XG 的工作范围为 550 mm,具有高精度、高速、优异的维护性、超大的容许惯性力矩,如图 8-8 所示。它可以根据动作开始时机械臂的姿态和动作结束时机械臂的姿态,自动选择最佳加速度、减速度。因此,只需输入最初搬运质量,电动机最大转矩和减速器允许最大转矩就不会超出允许值,不论何时都可以发挥电动机的最大功率,保持高加减速。

YK800XG

大型机型 YK800XG 适用于大型、重型工件的组装与搬运,如图 8-9 所示。YK800XG 的位置检测器采用旋转变压器。该旋转变压器采用无电子零部件的简单但强大的结构,具有环境适应性强、故障率低的特点,不会像光学编码器那样,会因为电子零部件故障、磁盘结露、黏附油污等因素而导致检测故障。而且,由于绝对式规格和增量式规格均为相同的机械规格,采用通用的控制器,因此只需设定参数即可变更规格。在绝对数据备份电池完全耗尽时,还能以增量式规格运作,即使万一发生情况,也无须停止生产线。

YK400XR

新开发的 SCARA 机器人 YK400XR 甚至重新设计了传统模式的机械部件。该 SCARA 机器人提供了高质量和高性能,而且性价比高。YK400XR 的臂展为 400 mm,有效负载为 5 kg,实现了标准周期时间 0.45 s,如图 8-10 所示。与新型控制器 RCX340 配套组合,可以实现各种高功能控制。

第 3 部分 水平关节机器人的结构组成

SCARA 机器人一般由三部分组成:操作机、控制器和示教器,如图 8-11 所示。

操作机

操作机又称机器人本体,是工业机器人的机械主体,用来完成规定任务的执行机构。主要由机械臂、驱动装置、传动装置和内部传感器组成。对 SCARA 机器人而言,其机械臂主要包括基座、大臂和小臂,如图 8-12 所示。

控制器

本节介绍爱普生 SCARA 机器人的 RC90 控制器和 YAMAHA SCARA 机器人的 RCX340 控制器。

1. RC90 控制器

爱普生 RC90 控制器是一款功能强大、价格低廉的控制器,适用于爱普生 LS 系列

SCARA 机器人,如图 8-13 所示。与爱普生其他控制器产品一样,RC90 控制器提供易用性的终极体验,同时价格降低了很多。RC90 可控制爱普生 LS3 和 LS6 SCARA 机器人,并提供卓越的 PowerDrive 伺服控制,实现平稳移动、快速的加减速速度和循环时间。爱普生 RC90 控制器具有较高的性价比。

RC90 控制器配备了爱普生 RC+Controls 软件和许多完全集成的选件。爱普生 RC+ 具有业界领先的易用性,与同类竞争机器人系统相比,爱普生 RC+ 向导、点击设置、进给和示教窗口、集成调试器、爱普生 Smartsense 及许多其他功能有助于降低整体开发时间。此外,爱普生 RC+ 开发环境集 Vision Guidance、.NET 支持、GUI Builder、DeviceNet、Profibus 等选件于一身,可实现最大性能和易用性。

2. RCX340 控制器

为了进一步提高 RCX 控制器的功能,YAMAHA 改善了所有功能,成功推出 RCX 控制器第三代产品 RCX340,如图 8-14 所示。

RCX340 控制器实现了控制器之间的高速通信。由于可以从主控制器向各从属控制器发出动作命令,程序、点位只需使用上级主控制器管理即可。此外,由于还可以灵活应对多任务,因此可以简化 PLC 的操作。各机器人可以同时启动、同时到达,自由控制,可以更容易、以更低的成本构建使用多个轴的复杂且精密的机器人系统。通过配备新伺服运动引擎,可以进行各种动作的连接。使用新开发的算法,成功缩短了定位时间并提高了轨迹精度。与以往的四轴控制器相比,体积约减小了 85%,实现了小型化,更容易设置在控制盘内。RCX340 控制器标配 RS232C 和 Ethernet 板。选配功能可支持 CC-Link、DeviceNet,以及 EtherNet/IP 等高速、大容量的各种网络。可以简单地连接其他公司的控制器以及其他公司的 VISION 视觉,RCX340 堪称"连接控制器"。

示教器

示教器是工业机器人的人机交互接口,机器人的绝大部分操作均可以通过示教器来完成。但对于 SCARA 机器人而言,有的厂商的示教器是作为选配件,如爱普生、YAMAHA 等,它们可以通过对应的软件来实现机器人控制。因此对 SCARA 机器人示教器不作介绍。

第 4 部分 水平关节机器人的应用

目前,SCARA 机器人非常适合医疗、汽车、电子、食品、实验室自动化、半导体、塑料、电器、航空航天等行业。它们可以用于各种各样的应用程序,例如零件的搬运和装配。

1. 搬运

SCARA 机器人的特点是其串接的两杆结构,类似人的手臂,可以伸进有限空间中作业然后收回,适合于搬运和取放物件,如集成电路板等,如图 8-15 所示。

2. 装配

SCARA 机器人在 X、Y 方向上具有顺从性,而在 Z 轴方向具有良好的刚度,此特性特别适合于装配工作。SCARA 机器人最先大量用于装配印刷电路板和电子零部件,如图 8-16 所示。

工业机器人技能考核实训台(HRG-HD1XKS)

HRG-HD1XKS 型 SCARA 机器人技能考核实训台,如图 8-17 所示,是一款通用型 SCARA 机器人实训台,以工业常用的 SCARA 机器人为核心,配备丰富的标准化工业应用教学模块,可以进行工业机器人虚拟仿真教学及 PLC、机器人编程等教学,可实现机器人视觉、物件装配、搬运、码垛等实训功能,主要用于SCARA机器人技术人才的培养教学和技能考核。

搬运模块

该模块可以进行机器人搬运应用学习,将托盘上的工件、物料从一个工位搬运到另一个工位。可用于学习机器人直线运动、圆弧运动、曲线运动的编程技巧,如图8-18所示。

第9单元 机器人的行业应用

第1部分 喷涂机器人

如果您的公司有喷涂应用,为什么不用机器人自动化?喷涂自动化是一种比任何手工喷涂过程都更简单、更安全和更优越的方法。此外,工业喷涂机器人比以前更容易获得,市场上不仅有更多的型号,而且比以往更加实惠。

所有的都用自动化,那会怎么样呢?这关系到做这个的成本和如何降低成本。那么,如果你正在喷涂,降低成本的方式是自动化!

时间成本

"时间就是金钱"一词绝对适用于喷涂和涂装工作。手工喷涂花费更多,因为需要更长的时间。一个工人不仅比一个喷涂机器人干得更慢,而且他们还不得不休息、就餐和休假。喷涂任务的重复性也可能导致疲劳、压力和伤害。另一方面,机器人能够每天工作24小时,每年365天。无论运行多长时间,他们都能够高效工作,提高工作量的同时不降低质量。

原料成本

当工作人员手工喷涂和涂装时,像过喷这种错误会浪费材料,降低产品质量,甚至可能损坏产品。手工喷涂的质量从来都是不一致的。总的来说,这是一个麻烦的过程,最终导致您的公司花费更多的钱。另一方面,机器人在保持油漆和工作方面具有令人难以置信的精度和一致性。使用机器人自动化时,典型的油漆节省率为15%~30%。由于它们被编程为在每个产品上喷洒相同数量的材料,所以它们很少有过喷问题,并且每次在每个部件上形成一致的涂层。

安全成本

油漆含有有害物质,如二甲苯和甲苯。这些物质在不同的喷涂和涂层材料中是常见的,并且由于化学品产生的有毒烟雾,人类长期在其中工作是非常危险的。相反,您可以选择自动化喷涂机器人,它可以控制和隔离上述危险。工人从危险中脱离出来,担任监督职务。

什么是喷涂机器人?

"喷涂机器人"是机器人的行业术语,与其他标准工业机器人有两个主要区别。

(1)防爆手臂。喷涂机器人是用防爆机器人手臂制造的,这意味着以这种方式制造的机器人可以安全地喷涂那些产生可燃气体的涂料。通常这些涂料是溶剂型涂料,喷涂时,需要建设一个消防安全监测的环境。

(2)独立涂料系统。当喷涂机器人首次设计时,它们只具有一个功能——在易挥发的环境中安全工作。随着接受和使用范围的扩大,喷涂机器人成为工业机器人的独特分支,而不仅仅是具有防爆功能的传统机器人。喷涂机器人现在有能力控制喷涂参数的各个方面。风机空气、雾化空气、流体流量、电压等都可以由机器人控制系统控制。

图 9-1 所示为 ABB 的喷涂机器人。

第 2 部分　焊接机器人

焊接机器人是从事焊接的工业机器人。工业机器人是一种多用途的、可重复编程的自动控制操作机,具有三个或更多可编程的轴,用于工业自动化领域。为了适应不同的用途,机器人最后一个轴的机械接口通常是一个连接法兰,可接装不同的工具或末端执行器。焊接机器人在工业机器人的末轴法兰上装接焊钳或焊(割)枪,使之能进行焊接、切割或热喷涂。

随着电子技术、计算机技术、数控及机器人技术的发展,自动焊机器人从 19 世纪 60 年代开始用于生产以来,其技术已日益成熟,主要有以下优点:

(1) 稳定和提高焊接质量,能将焊接质量以数值的形式反映出来;
(2) 提高劳动生产率;
(3) 改善工人劳动强度,可在有害环境下工作;
(4) 降低了对工人操作技术的要求;
(5) 缩短了产品改型换代的准备周期,减少了相应的设备投资。

因此,焊接机器人在各行各业已得到了广泛的应用。

焊接机器人主要包括机器人和焊接设备两部分。机器人由机器人本体和控制柜(硬件及软件)组成。而焊接装备,以弧焊为例,则由焊接电源(包括其控制系统)、送丝机、焊枪和其他部分组成。智能机器人还应有传感系统,如激光或相机传感器及其控制装置等。

另外,如果工件在整个焊接过程中无需变位,就可以用夹具把工件定位在工作台面上,这种系统是最简单的。但在实际生产中,多数的工件在焊接时需要变位,使焊缝处在较好的位置(姿态)下焊接。对于这种情况,定位机与机器人可以分别运动,即变位机变位后机器人再焊接;也可以是变位机变位的同时机器人进行焊接,也就是常说的变位机与机器人协调运动。这时定位机的运动及机器人的运动复合,使焊枪相对于工件的运动既能满足焊缝轨迹又能满足焊接速度及焊枪姿态的要求。实际上,这时定位机的轴已成为机器人的组成部分,这种焊接机器人系统可以多达 7~20 个轴或更多。

第 3 部分　搬运机器人

搬运机器人是可以进行自动化搬运作业的工业机器人。最早的搬运机器人于 1960 年出现在美国,Versatran 和 Unimate 两种机器人首次用于搬运作业。搬运作业是指用一种设备握持工件,将其从一个加工位置移到另一个加工位置。搬运机器人可安装不同的末端执行器,以完成各种不同形状和状态的工件搬运工作,大大减轻了人类繁重的体力劳动。目前世界上使用的搬运机器人超过 10 万台,被广泛应用于自动装配流水线、码垛搬运、集装箱等的自动搬运。部分发达国家已制定出人工搬运的最大限度,超过限度的必须由搬运机器人来完成。

搬运机器人是近代自动控制领域出现的一项高新技术,涉及了力学、机械学、液压气压技术、自动控制技术、传感器技术、单片机技术和计算机技术等学科领域,已成为现代机械制造生产体系中的一项重要组成部分。它的优点是可以通过编程完成各种预期的任务,在自身结构和性能上有人和机器的各自优势,尤其体现了人工智能和适应性。

常见的搬运机器人有串联关节机器人、水平关节机器人(SCARA 机器人)、Delta 并联关节机器人和 AGV 搬运机器人等。以上各类机器人在搬运行业中各有特点。

串联关节机器人拥有四个或六个旋转轴,类似于人类的手臂,更加灵活,负载较大,多用于较重物品的搬运。

SCARA 机器人是一种水平多关节机器人,工作频率较快,特别适合实验室自动化、医药、消费电子、食品、汽车、电子配件、PC 外设、半导体、塑料、家电和航空航天工业中物品和工件的搬运。

Delta 机器人属于高速、轻载的并联机器人。Delta 机器人主要应用于食品、药品和电子产品等的搬运、装配。Delta 机器人以其重量轻、体积小、运动速度快、定位精确、成本低、效率高等特点,在市场上广泛应用。

中国正处于产业转型升级的关键时刻,越来越多的企业在生产制造过程中引入工业机器人,它将深远地影响中国制造的方方面面。

第4部分 装配机器人

装配机器人是指工业自动化生产中用于对装配生产线上的零件或部件进行装配的一类工业机器人,是柔性自动化装配系统的核心设备。它用于精益工业生产,并增强了制造领域的生产能力。装配机器人可以提高生产速度和一致性。机械臂末端的工具可以为每个装配机器人定制,以满足生产要求,并适用于不同的应用。

用工业机器人进行装配操作可以使整个装配过程更快,提高效率和准确性。

1. 机器人装配提供了许多好处

在装配自动化中,机器人可能配备视觉技术,以适应不同的部件。另外,机器人视觉也可以提高效率和准确性。机器人可以执行单调乏味的组装任务,让工厂的人员去做其他的工作,同时提高质量。

2. 昼夜不停地生产是一个符合成本效益的选择

机器人一年到头每天 24 小时工作,不需要休息。它们将消除停工时间,降低劳动力成本,并提供高投资回报。

装配机器人广泛应用于各种电器制造、汽车及其部件、计算机、医疗、食品、太阳能、玩具、机电产品及其组件等行业,如图 9-2 所示。

装配机器人的分类

目前工业生产应用中,常见的装配机器人有 4 种类型:笛卡尔坐标机器人、六轴机器人、SCARA 机器人和 Delta 并联机器人。

笛卡尔坐标机器人主要应用于节能灯装配、电子类产品装配和液晶屏装配等场合;六轴机器人的应用范围最广,它能够适应绝大多数场合;SCARA 机器人广泛应用于电子、机械和轻工业等产品的装配;Delta 并联机器人主要应用于 IT、电子装配等领域。

装配机器人系统的组成

装配机器人系统主要由操作机、控制器、示教器、装配作业系统和周边设备组成,如图 9-3 所示。

1.装配作业系统

装配作业系统主要由搬运型末端执行器和真空负压站组成,操作机自带视觉系统。

2.周边设备

周边设备包括安全保护装置、机器人安装平台、输送装置、工件摆放装置、零件供给器等,用以辅助搬运机器人系统完成整个装配作业。

第5部分 打磨机器人

打磨机器人是指可进行自动打磨的工业机器人,广泛应用于3C、卫浴五金、IT、汽车零部件、工业零件、医疗器械、民用产品等行业,如图9-4所示。

机器人打磨是精炼表面的过程,直到表面光滑发亮。这个应用程序是重复且单调的,需要极端的一致性。为了得到一致、精湛、高质量的产品,打磨机器人被设定为可提供适当的压力,并向正确的方向精确地移动。

1.机器人去除材料的优点

具有灵活性、可重复性和极高的精确度的打磨机器人可以研磨、修整或抛光几乎任何材料,以生产一致的、高质量的成品。这些机器人还可以在减少浪费的同时改善生产时间。机器人也能让工人免于因打磨而带来的苦差事和安全隐患。打磨机器人不会受到烟雾和灰尘的伤害。此外,机器人打磨对环境更有利,因为干燥的研磨轮代替了化学溶液。

2.机器人处理过程的方法

在目前的实际应用中,打磨机器人大多数是六轴机器人。根据末端执行器性质的不同,打磨机器人分为两类:机器人持工件和机器人持工具,如图9-5所示。

机器人持工件通常用于需要处理的工件相对比较小时,机器人通过其末端执行器抓取待打磨工件并操作工件在打磨设备上打磨。另外,可以通过让机器人把成品放在传送带上或类似的装置上,来增加系统的价值。一般在该机器人的周围有一台或数台工具。然而,机器人持工具一般用于大型工件或对于机器人来说比较重的工件。工件的装卸可由人工进行,机器人自动地从工具架上更换所需的打磨工具。通常在此系统中采用力控制装置来保证打磨工具与工件之间的压力一致,补偿打磨头的损耗,获得均匀一致的打磨质量,同时也能简化示教。

在实际应用中,也有可能让几个机器人一起工作以获得最终的灵活性。一个机器人控制部件,其他的机器人操纵工具。

打磨机器人系统的组成

持工具的打磨机器人系统主要由操作机、控制器、示教器、打磨作业系统和周边设备组成,如图9-6所示。

1.打磨作业系统

打磨作业系统包括打磨动力头、变频器、力传感器、力传感器控制器和自动换刀装置。

2.周边设备

周边设备包括安全保护装置、机器人安装平台、输送装置、工件摆放装置、消音装置,用以辅助打磨机器人系统完成整个打磨作业。

第 10 单元　新型机器人

第 1 部分　YUMI

机器人交互合作的新时代已经到了。YuMi 是 ABB 多年研究与开发的结果，能使人与机器人的合作成为现实，而且还有更多其他意义。

ABB 开发了一种协同的、双臂的、用于小部件装配的机器人解决方案，它包括灵活的手、零件供料系统、基于相机的零件定位和最先进的机器人控制技术。YuMi 是一个未来的愿景。YuMi 将改变我们对装配自动化的思考方式。YuMi 是"你和我"，共同创造无限的可能性。

结构介绍

IRB 14000 是 ABB 机器人有限公司的第一代双七轴臂机器人，专为利用机器人的灵活自动化的制造行业（例如 3C 行业）而设计，如图 10-1 所示。该机器人为开放式结构，特别适合于灵活应用，并且可以与广泛的外部系统进行通信。

操作系统

该机器人配备有控制器（内置在机器人中）和机器人控制软件 RobotWare。RobotWare 支持该机器人系统的方方面面，如动作控制、应用程序的开发和执行以及通信等。

附加功能

关于附加功能，该机器人可以配备可选软件以获得应用程序支持，如通信功能以及多任务、传感器控制等高级功能。

机械臂的轴

YuMi 每个手臂有 7 个轴，多自由度使其像人手一样灵活，如图 10-2 所示。

机臂配置对双臂都适用。

安全功能

控制系统采用了固有设计的安全措施，拥有限制功率和力度。ABB 机器人的示教器如图 10-3 所示。

第 2 部分　SDA10F

SDA10F 是一款双臂、十五轴机器人，具有令人难以置信的灵巧性，运动灵活，占地面积小，如图 10-4 所示。两臂可以一起工作，大大简化了手臂末端工具。采用专利伺服执行机构设计，所有电缆均通过手臂布线。FS100 是具有无与伦比的开放软件架构的强大的控制器。

主要优点

具有执行复杂任务的灵巧性：七轴双臂可以一起工作或独立工作；纤细的设计优化了空间，即使在狭小的空间中也能提供"像人一样"的灵活性和运动范围；简化的工具降低了成本；可用于对人类有害的劳动环境中，节省资本投入。

纤细，双臂机器人具有"类人"的灵活性。

(1) 卓越的灵巧性和一流的手腕特性使纤细的双臂机器人完美地适用于以前只能由人完成的组装、零件转移、机器管理、包装和其他处理任务。

(2) 高度灵活：15 个运动轴(每个臂 7 个轴,加上单个轴用于基座旋转)。

(3) 强大的基于执行器的设计,能提供"类人"的灵活性,可快速加速。

(4) 内部布线的电缆和软管(6 个空气的,12 个电气的)可减少干扰和维护,并且使编程更容易。

(5) 每臂有效负载 10 kg；每臂水平伸展距离 720 mm；每臂垂直伸展距离 1 440 mm；重复精度±0.1 mm。

(6) 机器人的两个手臂可以协同工作,执行同一个任务,使有效负载翻倍,或者搬运沉重、笨重的物体；两臂可以同时进行独立操作。

(7) 在进行附加操作时可以用一只手握住部件,不需要将零件放下,另一个手臂将零件从一个手臂转移到另一个臂。

第 3 部分　Baxter

数十年来,制造商对处理低容量、高混合生产工作的成本效益非常低。于是,他们遇到了 Baxter——刚性自动化中安全、灵活、经济的选择。北美的领先公司已经把 Baxter 纳入员工队伍,并获得了竞争优势。

Baxter(见图 10-5)是一个经过验证的工业自动化解决方案,可以处理广泛的任务,如从线路加载和机器管理,到包装和材料搬运。如果您走到您的设备旁,看到有单调或危险的工作任务,那么 Baxter 就可以为您的公司工作了,它能很准确地做单调的工作,从而解放你的技术工人。

Baxter 的主要优点如下。

(1) 设计安全：Baxter 可以安全地在生产环境中操作,不需要围栏,可以节省金钱和宝贵的场地。

(2) 容易集成：Baxter 可以快速部署,无缝连接到其他自动化设备上,通常无需第三方集成。

(3) 示教,不用编程：使用 Baxter 不需要传统的编程。相反,它通常由内部员工进行示教,从而减少第三方程序员参与的时间和成本。

(4) 自适应性：Baxter 的兼容性手臂和力传感器使其能够适应可变的环境,"感觉"异常和引导部件到位。

(5) 灵活性和可重复部署性：Baxter 可灵活运用于各种应用,并可跨线路和任务重新训练。Baxter 能重新快速适应工作,通常一年就能收回投资。

(6) 经济实惠和可扩展：Baxter 的基础价格为 2.5 万美元,对于中小企业来说是可行的,其性能可以通过改进常规软件不断改进。

Baxter 的应用包括但不限于搭配、打包、上料和下料、机器管理、材料搬运。真空夹具的附件可以拾取多个种类的物体,特别是光滑的、无孔的或平坦的物体；电动平行夹爪可用于拾取多种形状和尺寸的刚性或半刚性物体；可互换的手指和指尖使灵活性最大化；使用工业级脚轮的移动式底座可以方便地在工作站之间快速安全地移动 Baxter。

Baxter 由称为"Intera"的平台驱动,发音为"in-terra",命名反映了机器人的交互式生产功能。Intera 提供了易于使用的图形用户界面,内部员工可以快速掌握。该平台允许

Baxter 通过演示进行训练，使用文本而不是坐标使非技术人员可以根据需要创建和修改程序。

它智能地处理不断变化的环境，同时提供一个可扩展的平台，利用现代工具（如 ROS）来最大限度地发挥现代劳动力的相关性和劳动力。

第 4 部分　YouBot

它是小巧、有趣、多功能，为未来的发明者而制作，KUKA youBot 是一个强大的教育机器人，特别设计用于移动操纵的研究和教育，这是专业服务机器人的关键技术，如图 10-6 所示。

KUKA youBot 由两个主要部分组成，包括一个全方位的平台，一个五自由度的机器人手臂与双手指夹。

KUKA youBot 全方位移动平台由机器人底盘、四个麦克风轮、电动机、电源和板载 PCC 板组成，如图 10-7 所示。用户可以在该板上运行程序，也可以在远程计算机中进行控制。该平台配有一个带有预安装 Ubuntu Linux 的 Live-U 盘以及硬件驱动程序。

KUKA youBot 手臂有五个自由度（DOF）和一个双手指夹，如图 10-8 所示。如果连接到移动平台，手臂可以由板载 PC 控制，也通过以太网电缆连接到自己的 PC，不通过移动平台进行控制。

机器人上可以安装额外的传感器。

KUKA youBot 具有完全开放的接口，允许开发人员在几乎所有级别的硬件控制上访问系统。还配备一个应用程序编程接口（KUKA youBot API）以及最新机器人框架（如 ROS 或 ORCOS）的接口与封装，具有 Gazebo 开源仿真和一些示例代码演示如何编程 KUKA youBot。该平台和可用的软件将使用户自己能够快速开发移动操纵应用程序。

第 5 部分　NAO

NAO 是一个自主的、可编程的类人机器人，如图 10-9 所示，由 Aldebaran 机器人公司开发。该公司是一个公司总部位于巴黎的法国机器人公司。机器人的发展始于 2004 年发起的 NAO 项目。2007 年 8 月 15 日，NAO 取代索尼的机器狗"爱宝"成为"机器人足球世界杯"（RoboCup）标准平台联盟（SPL）中使用的机器人。机器人足球世界杯是一场国际机器人足球竞赛。NAO 被用于 2008 年和 2009 年举办的 RoboCup，2010 年的 RoboCup 中，NAOV3R 被选为 SPL 的比赛平台。

自 2008 年以来，NAO 机器人已经发布了多个版本。NAO 学术版是以学校和实验室的研究和教育为目的开发的。NAO 机器人在机器人技术、系统、控制、计算机科学、社会科学和其他方面是开创性的辅助教学方法。类人机器人能吸引人类尤其是学生。NAO 机器人允许他们开发程序、传感器，也可以与人和环境互动，还有其他功能。NAO 机器人在全球多个学术机构被用于研究和教育，截至 2015 年，在 50 多个国家有超过 5 000 个 NAO 机器人被使用。

NAO 能够自主移动，与人类交谈，识别对象并与其环境进行交互。任何人都可以轻松地通过 Choregraphe 软件的图形界面编写程序。学生可以使用可配置的行为框来开发基于事件的、顺序的或并行的程序。用户还可以创建自己的行为，以及使用 Python 编写更复杂的脚本。

NAO可以通过以太网电缆或WiFi连接机器人和计算机进行编程。NAO的软件Webots允许用户在具有现实世界物理现象的虚拟环境中模拟机器人程序。屏幕上的NAO可以与虚拟世界中的对象交互,允许学生测试他们的程序,同时其他人也可以使用NAO的硬件。该软件套件与Windows、Mac和Linux兼容。

第11单元 工业机器人展望

第1部分 工业机器人现状

机器人是机械、电子、控制、传感、人工智能等多学科的先进技术相结合的自动化设备之一。全球机器人的发展现状如下。

1. 全球机器人市场需求持续增长

工业机器人和服务机器人的市场规模持续扩大。根据IFR(国际机器人联合会)的统计,2015年全球工业机器人销量首次突破24万台,其中亚洲销量约占全球销量的2/3,销量为14.4万台;欧洲地区为5万台,其中东欧地区销量增速达到29%,是全球增长最快的地区之一;北美地区销量达到3.4万台,较2014年同比增长11%。中国、韩国、日本、美国和德国的总销量占全球销量的3/4。中国、美国、韩国、日本、德国、以色列等国是近年工业机器人技术、标准及市场发展较活跃的地区。近年来,全球工业机器人销量处于稳步增长态势。

2. 亚太地区成为最重要市场

根据IFR的统计,亚洲是目前全球工业机器人使用量最大的地区,占世界范围内机器人使用量的50%,其次是美洲(包括北美、南美)和欧洲。2012—2015年亚洲机器人销量年均增长15%,远高于美洲和非洲6%的增长速度。2015年,亚太地区工业机器人销量超过14万台。2014年中国、日本、韩国和泰国的工业机器人新装机量占亚洲地区总量的75%,这四个国家工业机器人的市场规模占全球工业机器人销量的52.4%。

3. 工业机器人发展高度集中

工业机器人的主要产销国集中在日本、韩国和德国,这三国的机器人保有量和年度新增量位居全球前列。

日本、韩国和德国的机器人密度和保有量处于全球领先水平。据IFR统计,2014年日本每万名工人拥有323台工业机器人,韩国为437台,德国为282台;2013年日本的机器人保有量为30.4万台,韩国为15.6万台,德国为16.8万台。

2014年,日本、韩国、德国三国的机器人市场新增量占全球的30.9%,市场规模分别为2.9万台、2.1万台、2万台。日本机器人市场成熟,其制造商国际竞争力强,FANUC、NACHI、Kawasaki等品牌在微电子技术、电力电子技术领域持续领先。韩国的半导体、传感器、自动化生产等高端技术为机器人的快速发展奠定了基础。德国工业机器人在人机交互、机器视觉、机器互联等领域处于领先水平,德国本土的KUKA公司是世界工业机器人四大制造商之一,年产量超过1.8万台。

4. 服务机器人市场处于起步阶段

全球服务机器人市场仍然处于起步阶段。一是由于服务机器人的外围技术问题未能解决。服务机器人技术是多学科交叉集成技术，涉及机械设计、自动控制、仿生学、运动学等多领域，在多样性、随机性、复杂性的环境背景下，其对于环境感知的任务复杂度和实时性要求更高。二是单位价值高的服务机器人整体水平技术低下，发展速度缓慢。如医用机器人的控制运动、精细组织操作和三维高清晰度的视觉能力要求高，仅少量发达国家有能力采用此类技术。

目前全球服务机器人市场仅有部分国防机器人、家用清洁机器人、农业机器人实现了产业化，而技术含量更高的医疗机器人、康复机器人等仍然处于研发试验阶段。全球个人和家用服务机器人的产品包括家庭作业机器人、娱乐休闲机器人、残障辅助机器人和监视机器人，其中家庭作业机器人中的除草机器人市场化程度高，产品种类多样化。例如，达芬奇手术机器人、挤奶机器人和军用无人机已经形成成熟的产业链。

第2部分　机器人发展趋势

1. 机器人与信息技术深入融合

大数据和云存储技术使得机器人逐步成为物联网的终端和节点。一是信息技术的快速发展将工业机器人与网络融合，组成复杂性强的生产系统，各种算法如蚁群算法、免疫算法等可以逐步应用于机器人应用中，使其具有类人的学习能力，多台机器人协同技术使一套生产解决方案成为可能。二是服务机器人普遍能够通过网络实现远程监控，多台机器人能提供流程更多、操作更复杂的服务；人类意识控制机器人这一新操作模式也正在研发中，即利用"思维力"和"意志力"控制机器人的行为。

2. 机器人产品易用性与稳定性提升

随着机器人标准化结构、集成一体化关节、自组装与自修复等技术的改善，机器人的易用性与稳定性不断提高。一是机器人的应用领域已经从较为成熟的汽车、电子产业延伸至食品、医疗、化工等更广泛的制造领域，服务领域和服务对象不断增加，机器人本体向体积小、应用广的特点发展。二是机器人成本快速下降。机器人技术和工艺日趋成熟，机器人初期投资相较于传统专用设备的价格差距缩小，在个性化程度高、工艺和流程烦琐的产品制造中具有更高的经济效益。三是人机关系发生了深刻改变。例如，工人和机器人共同工作时，机器人能够通过简易的感应方式理解人类语言、图形、身体指令，利用其模块化的插头和生产组件，免除工人复杂的操作。现有阶段的人机协作存在较大的安全问题，尽管具有视觉和先进传感器的轻型工业机器人已经被开发出来，但是目前仍然缺乏安全可靠的工业机器人协作的技术规范。

3. 机器人向模块化、智能化和系统化方向发展

目前全球推出的机器人产品向模块化、智能化和系统化方向发展。第一，模块化改变了传统机器人结构仅能适应有限范围的问题，工业机器人的研发更趋向采用组合式、模块化的产品设计思路，重构模块化帮助用户解决产品品种、规格与设计制造周期和生产成本之间的矛盾。第二，机器人产品向智能化发展的过程中，工业机器人控制系统向开放性控制系统集成方向发展，伺服驱动技术向非结构化、多移动机器人系统改变，机器人协作已经不仅是控制的协调，而且是机器人系统的组织与控制方式的协调。第三，工业机器人技术不断延伸，目前的机器人产品正嵌入工程机械、食品机械、实验设备、医疗器械等传统装备之中。

4. 新型智能机器人市场需求增加

新型智能机器人，尤其是具有智能性、灵活性、合作性和适应性的机器人的需求持续增

长。第一，下一代智能机器人的精细作业能力进一步提升，对外界的适应感知能力不断增强。在机器人精细作业能力方面，波士顿咨询集团调查显示，最近进入工厂和实验室的机器人具有明显不同的特质，它们能够完成精细化的工作内容，如组装微小的零部件，预先设定程序的机器人不再需要专家的监控。第二，市场对机器人灵活性方面的需求不断提高。雷诺目前使用了一批 29 kg 的拧螺丝机器人，它们在仅有的 1.3 m 长的机械臂中嵌入 6 个旋转接头，机械臂能灵活操作。第三，机器人与人协作能力的要求不断增强。未来机器人能够靠近工人执行任务，新一代智能机器人采用声呐、摄像头或者其他技术感知工作环境是否有人，如有碰撞可能，它们会减慢速度或者停止运作。

第 3 部分　应用机器人引发的社会问题

目前大多数机器人主要应用于工业领域，但针对人们日常生活需求，各种机器人已经逐渐向我们走来。在机器人应用越来越广泛的时代背景下，机器人可能会深刻地改变我们的生活方式，由此引发的伦理问题也日益紧迫地摆在我们面前。近十余年来，机器人伦理研究也是学术界的一个热门话题。

1. 军用机器人：是爱惜生命还是无情杀戮

机器人在军事方面的运用是机器人技术发展进步的重要推动力量之一。目前，世界各国都在积极研发包括地面机器人、空中机器人、水下机器人以及空间机器人等各式各样的军用机器人。与人类士兵相比，军用机器人可以在更加恶劣的环境中作战，而且绝对服从命令，也不需要反复训练，其优势显而易见。让军用机器人代替人类士兵在战场上冲锋陷阵，可以很大程度地减少人员的伤亡。因此，军用机器人的使用似乎是人类珍爱生命的一种重要手段。

但是，在现代科学技术武装下的军用机器人可以拥有比人类士兵更加强大的破坏力。更何况，机器人可能对人类没有同情心，它们拥有的强大杀伤力可以使其成为真正的冷血"杀人机器"。当然，人们可以通过约定"机器人与机器人作战、机器人不与人类作战"的方式来解决这一问题，但这种解决方式很难实现。因此，关于军用机器人的伦理问题是机器人伦理研究中讨论最多的内容之一。

2. 儿童看护机器人：减轻负担还是推卸责任

儿童看护机器人在韩国、日本和少数欧洲国家得到广泛重视。儿童看护机器人具备视频游戏、语音识别、面部识别以及会话交流等多种功能。它们装有视觉和听觉监视器，可以移动，自行处理一些问题，在孩子离开规定范围时还会报警。

儿童看护机器人会受到孩子的喜欢吗？实验表明，儿童看护机器人可以作为幼儿的亲密玩伴，特别是对缺少小伙伴的幼儿来说可能更重要，因为它们的效果比一般的玩具要好得多，可以让他们更开心。儿童看护机器人已经越来越多地进入普通家庭。

毫无疑问，儿童看护机器人可以减轻家长负担，使他们能有更多的自由时间。但是，对儿童的健康成长来说，家长的关爱是无法替代的，机器人只能起到辅助作用。如果把儿童较多地交给机器人照顾，可能会影响其心理和情感的正常发展。

3. 机器人陪伴：排解孤独还是脱离社会

人口老龄化是世界许多国家人口发展的普遍规律，我国亦不例外。从单个家庭的角度看，助老机器人的协助可以减少家人照顾老人的时间，从而减轻家庭负担。机器人代替人类护工，可以减轻人类护工的工作量，减少社会对人类护工的需求量，从而减轻社会负担。而且,与人类护工相比,助老机器人还有一些独特的优点。比如,机器人可以 24 小时为老人服务。又如,各种各样的机器人可以满足老人不同的需要,显著提高老人的生活质量,甚至还可以根据老人的需要定制机器人,人类护工则难以做到这一点。另外,助老机器人还可以与老人互动、娱乐,甚至在一定程度上可以满足老人的情感需求,使独立生活的老人减轻孤独感。大家对助老机器人的使用几乎都持有积极和乐观的态度。

但是,助老机器人也可能引发一系列伦理问题,比如老人的隐私、自由与尊严等问题,其中最突出的可能是老人与家人、社会联系的问题。如果把许多本应由人类完成的工作全部交给机器人来做,有可能大大降低老人与社会的交流与联系。研究表明,较多的社会交流与互动,可以延长那些需要长期护理的老人的寿命。毋庸置疑,老人也有较强的情感需要,这种需要是否得到满足,以及在多大程度上得到满足,对于老人的健康至关重要。

4. 机器人技术及其应用:自由发展,还是伦理规制

有的科学家认为科学技术是价值中性的,认为科学技术研究无禁区,技术应用导致的后果跟技术本身无关。怎么用它,究竟是给人类带来幸福还是带来灾难,全取决于人类自己,而不取决于工具。也有学者认为,现有的关于机器人的伦理考量主要是建立在理论研究的基础上,现实情况将会如何,尚不能定论,我们应该先将相关的科学技术发展起来再说,不能因为一些可能产生的负面效应而放缓步伐。如果人文社会科学的研究主要依据已有的技术充分发展及其社会效应充分显现之后,就会产生"文化之后现象"和"制度真空",这对人类文明的发展显然是不利的。因此,人文社会科学研究必须具有一定的预测性和前瞻性,以防患于未然。

鉴于机器人技术可能产生的深刻社会影响,对其进行伦理规制与重视技术发展同样重要。机器人技术的伦理规制就是将机器人伦理原则规范化、制度化、具体化,其目的是最大程度地保护人类整体与个体的利益。

如果说 21 世纪将发生影响深远的"机器人革命"的话,那么这场革命将不仅是科学技术革命,也将是一场社会与伦理的革命。机器人的广泛使用是不可避免的,我们需要关注的不是该不该使用的问题,而是如何更好地使用的问题。为了使机器人更好地为社会服务,技术上的发展当然是必需的,同时伦理方面的考量也至关重要。

第 4 部分　工业机器人最新行业数据

目前,全球机器人市场规模持续扩大,工业、特种机器人市场增速稳定。技术创新围绕仿生结构、人工智能和人机协作不断深入,产品在教育陪护、医疗康复、危险环境等领域的应用持续拓展,企业前瞻布局和投资并购异常活跃,全球机器人产业正迎来新一轮增长。

根据《2018 年世界机器人报告》,2017 年全球工业机器人销售量创下 38.7 万台新纪录,与上一年相比增长了 31%(2016 年:29.43 万台)。中国对工业机器人的需求增长最快,增幅达 58%。与前一年相比,美国的销量增长 6%,德国增长 8%。

按产业来看,汽车业继续引领全球对工业机器人的需求:2017 年对该产业售出约 12.55 万台工业机器人,相当于增长 21%。2017 年增长最强劲的产业分别为金属业(+55%)、电气/电子业(+33%)和食品业(+19%)。

在销量方面,亚洲拥有最广阔的个人市场:2017 年中国安装了约 13.8 万台工业机器人,其后依次为日本,约 4.6 万台,韩国约 4 万台。在美洲,美国是最大的单一市场,其工业机器人销量约 3.3 万台,而德国是欧洲最大的单一市场,销量约 2.2 万台。

国际机器人联合会会长 JunjiTsuda 表示:"全球工业机器人继续以令人印象深刻的速度增长。数码化、简单化和人类-机器人协作等关键趋势必将塑造未来并推动其快速发展。"

在数码化过程中,实际生产与虚拟资料世界的联系越来越密切,开启了从分析直至机器学习的全新可能性。机器人将能通过学习过程而获得新技能。与此同时,业界正在努力简化机器人的使用。在未来,利用直观的程序将使工业机器人程式设计更加容易、更加快速。

第 12 单元　智能制造与全球机器人发展计划

第 1 部分　工业 4.0

第四次工业革命,也被称为工业 4.0、未来工厂、智慧工厂、工业互联网,正在迅速涌现,并以新的方式影响我们的生活。它有助于简化机器人及其自动化操作,同时还可以优化成本,将企业广泛的自动化业务转型的潜力变成现实。

第四次工业革命是自动化和数据交换的新融合,为"智慧工厂"创造了完美的环境。它是一种将技术领域中所有的进步都结合和联网的新的连接:智能系统(网络物理系统)、物联网和云计算。所有的物品被编程为一起工作,使人和机器能够实时甚至远程地进行协作和通信。生产链中所有设备、系统和人员都被连接到一起,并能够随时随地地以正确的形式传递数据。

工业 4.0 正在不断地证明,人类正在不可阻挡地向着更加有效和智慧的社会方向创新。从一开始,人类就一直不断地研究,不断地创新和寻找新的解决方案,使创造产品更实惠、更智能,也使我们的生活更容易一些。过去的革命(水和蒸汽动力、电力和数字动力)真正改变了社会、管理结构和人的特性。所有这些革命都帮助制造业不断寻求更好、更经济可行的解决方案。

第一次工业革命开始使用化石燃料和机械动力作为能源。第二次工业革命带来电力分配、有线和无线以及新形式的发电。第三次革命带来了计算能力的快速发展,实现了新的生成、处理和共享信息的方式。它还使用电子和 IT 系统来进一步实现制造的自动化。所有这些革命都不仅由研究人员、发明家和设计者驱动,更重要的是在日常生活中采用和使用这些革命的人。工业 4.0 的到来真正反映了人们的愿望和选择。

四次工业革命一览图如图 11-1 所示。

所以我们今天看到的新的工业革命真的没有什么奇怪的。第四次工业革命依靠现有的所有革命,为人和机器提供全新的功能,以及技术在社会中的嵌入方式。制造敏捷性的水平可以使客户需求与公司提供产品的能力、潜在的需求相关联。制造商能够倾听,按需访问实时情报和分析,然后采取行动更好地适应消费者的需求。

如果我们有勇气为正在进行的变革承担主要责任,共同努力提高认识和形成新的规则,我们就可以重新调整经济、社会和政治制度,充分利用新兴技术。

很明显地,第四次工业革命已经到来,如果你在制造业,袖手旁观不再是一个好主意。机器人技术正在彻底改变企业成功的方式,其利益越来越明显,可达到各种规模。

第 2 部分　智能制造的核心技术

智能制造（intelligent manufacturing，IM）是指由智能机器和人类专家共同组成的人机一体化智能系统。它在制造过程中能进行智能活动，诸如分析、推理、判断、构思和决策等，通过人与人、人与机器、机器与机器之间的协同，去扩大、延伸和部分地取代人类专家在制造过程中的脑力劳动。

智能制造是机械化、自动化和信息化应用到成熟阶段的必然产物。

智能制造包括以下核心技术。

1. 信息物理系统

信息物理系统（CPS）基本上是集成智能、人机网络互联的统称。工厂管理者不是简单地重组生产线，而是积极创建一个机器人网络，这样在保持高效率的同时，不仅能够减少失误，还能够根据外部投入自主地改变生产模式。

2. 人工智能

AI（Artificial Intelligence）：即人工智能，它是研究、开发用于模拟、延伸和扩展人的智能的理论、方法、技术及应用系统。它企图了解智能的实质，并生产出一种新的、能以和人类智能相似的方式做出反应的智能机器，该领域的研究包括机器人、语言识别、图像识别、自然语言处理和专家系统等。

3. 增强现实技术

它使用技术手段在我们看到的世界上叠加信息。例如，在用户看到和听到的内容上附加图像和声音。想象一下《少数派报告》和《钢铁侠》的互动风格，这与虚拟现实有很大的不同。虚拟现实意味着你沉浸在计算机生成的环境中并与之进行互动。增强现实（也称为AR），增加了你通常所看到的现实性，而不是取代它。增强现实往往被视为一种未来技术，但宽泛一点讲，它已经以某种形式存在了多年。例如，早在 20 世纪 90 年代，许多战斗机中的平视显示器能显示飞机的姿态、方向和速度的信息，几年后，它们甚至能够显示视野中的哪些物体是目标。

4. 物联网

IoT（Internet of things）：即物联网，物联网就是物物相连的互联网，指通过各种信息传感设备，实时采集任何需要监控、连接、互动的物体等各种需要的信息，与互联网结合形成的一个巨大网络。其目的是实现物与物、物与人，所有的物品与网络的连接，方便识别、管理和控制。

5. 工业大数据

IBD（industrial big data）：即工业大数据，大数据理念已应用于工业领域，使设备数据、活动数据、环境数据、服务数据、经营数据、市场数据和上下游产业链数据等原本孤立、海量、多样性的数据相互连接，实现人与人、物与物、人与物之间的连接，尤其是实现终端用户与制造、服务过程的连接。通过新的处理模式，根据业务场景对实时性的要求，实现数据、信息与知识的相互转换，使其具有更强的决策力、洞察发现力和流程优化能力。相比其他领域的大数据，工业大数据具有更强的专业性、关联性、流程性、时序性和解析性等特点。

第 3 部分　智慧工厂

近年来，我国不断出台针对性举措推动智能制造的发展，为传统制造型厂商指出了新的发展方向——智慧工厂。智慧工厂的提出,将搭建起产品和制造之间的沟通桥梁,为智能制造起到承接落地的作用。因此,制造业的未来将是智慧工厂。

所谓智慧工厂,是在数字化工厂的基础上,集各种新兴技术和智能系统于一体而构建的人性化工厂。智慧工厂提高了生产过程可控性,减少了生产线人工干预,及时准确地采集作业数据,因此能增强核心竞争力、提高生产效率及合理安排生产等。

从定义可以看出,智慧工厂的实现离不开新兴技术的支持与应用。例如,没有先进传感器的广泛应用,智慧工厂难以称得上智能。而先进传感器的发展依赖于微处理器和人工智能技术的进步。

软件工程辅助系统的应用,也是智慧工厂的基本构成要素之一。软件工程辅助系统是基于知识的、高度集成的智能软件系统,拥有数字量表达、信息获取、知识处理等能力。智慧工厂将逐步取代传统、单一的工作模式。

当然,为真正实现智慧工厂,还需要重点突破机床、工艺、生产管控等关键技术。

以机床为例,数控机床是智慧工厂最基本的部分。因此,智慧工厂的建设离不开智慧机床的进步。智慧机床可以独立收集数据并确定运行状态,自动检测、诱导、模拟目标的智能决策,使机器运行处于最佳状态。

此外,智能物流的发展也是智慧工厂快速发展的关键因素。对于智慧工厂而言,智能物流是一个"回收系统",需要持续运输生产相关资源。幸运的是,中国近年来大力发展智能物流,信息化、智能化建设取得了长足进展。

总的来说,智慧工厂是现代工厂信息化发展的最终目标,也是实现智能制造的重要一步。

附录2　常用缩略词

3C	computer, communication, consumer electronic	计算机、通信、消费类电子产品
4D	four-dimensional	四维
ABB	ASEA Brown Boveri	艾波比集团公司
AGV	automated guided vehicle	自动导引运输车
AI	artificial intelligence	人工智能
AMF	American Machine and Foundry	美国机械与铸造公司
API	application program interface	应用程序接口
AR	augmented reality	增强现实
ARM	advanced robot motion	高级机器人运动
ASEA	Allmänna Svenska Elektriska Aktiebolaget (predecessor of ABB)	阿西亚公司（ABB前身）
ATC	automatic tool changer	自动工具交换装置
BBC	Brown, Boveri & Cie	布朗·勃法瑞公司
CC-Link	control and communication link	控制与通信链路系统
CCU	cabinet control unit	内控单元
CNC	computer numerical control	数控系统
CPS	cyber-physical system	信息物理系统
CSP	controller system panel	控制器系统面板
DC	direct current	直流
DOF	degrees of freedoms	自由度
EL	electro-luminescence	电发光
EoAT	end of arm tooling	手臂末端工具
FSU	functional safety unit	安全功能模块
GMAW	gas metal arc welding	熔化极气体保护焊
HMI	human machine interface	人机界面
HRC	human-robot collaboration	人类-机器人协作
I/F	current intensity/frequency	电流/频率转换
I/O	input/output	输入/输出端口
IBD	Industrial big data	工业大数据
IFR	Internation Federation of Robotics	国际机器人联合会
IM	intelligent manufacturing	智能制造工业
IoT	The Internet of things	物联网

IP	internet protocol	网络协议
IPA	International Profession Certification Association	国际认证协会
IPS	integrated paint system	综合喷绘系统
ISO	International Standards Organization	国际标准化组织
IT	Internet technology	IT 技术
KPC	control PC	控制计算机
LCD	liquid crystal display	液晶显示屏
NC	numerical control	数控
PDA	personal digital assistant	个人数字助理
PC	personal computer	个人电脑
PUMA	programmable universal machine for assembly	通用工业机器人
R.U.R	Rossum's Universal Robots	罗萨姆万能机器人
RIA	Robotics Institute of America	美国机器人学会
RS-232C	recommended standard	串行物理接口标准
SCARA	selective compliance articulated robot arm	选择顺应性关节机器手臂
SIB	safety interface board	连接安全信号的接口板
SME	small and medium enterprise	中小型企业
SPL	Standard Platform League	标准平台联赛
TCP	transmission control protocol	传输控制协议
US	the United States	美国

参 考 文 献

[1] 张明文. ABB六轴机器人入门实用教程[M]. 哈尔滨:哈尔滨工业大学出版社,2017.
[2] 张明文. 工业机器人技术基础及应用[M]. 哈尔滨:哈尔滨工业大学出版社,2017.
[3] http://new.abb.com.
[4] http://www.fanuc.co.jp/eindex.html.
[5] https://www.kuka.com.
[6] http://www.yaskawa.com.cn.
[7] https://en.wikipedia.org/wiki/Industrial_robot.
[8] https://www.robots.com/education/industrial-robot-history.
[9] 朱晓玲. 机电工程专业英语[M]. 北京:机械工业出版社,2007.
[10] 杨春生. 机电专业英语[M]. 北京:电子工业出版社,2007.

教学课件下载步骤

步骤一
登录"工业机器人教育网"
www.irobot-edu.com，菜单栏点击【学院】

步骤二
点击菜单栏【在线学堂】下方找到您需要的课程

步骤三
课程内视频下方点击【课件下载】

咨询与反馈

尊敬的读者

　　感谢您选用我们的教材！

　　本书配套有丰富的教学资源，凡使用本书作为教材的教师可咨询有关实训装备事宜，在使用过程中，如有任何疑问或建议，可通过邮件（edubot_zhang@126.com）或扫描右侧二维码，在线提交咨询信息，反馈建议或索取数字资源。

培训咨询：+86-18755130658（郑老师）
校企合作：+86-15252521235（俞老师）

（教学资源建议反馈表）